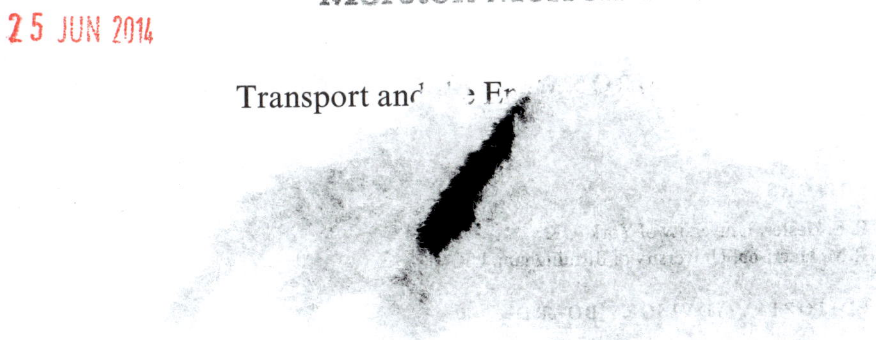

Transport and the En

ISSUES IN ENVIRONMENTAL SCIENCE AND TECHNOLOGY

TITLES IN THE SERIES:

1 Mining and its Environmental Impact
2 Waste Incineration and the Environment
3 Waste Treatment and Disposal
4 Volatile Organic Compounds in the Atmosphere
5 Agricultural Chemicals and the Environment
6 Chlorinated Organic Micropollutants
7 Contaminated Land and its Reclamation
8 Air Quality Management
9 Risk Assessment and Risk Management
10 Air Pollution and Health
11 Environmental Impact of Power Generation
12 Endocrine Disrupting Chemicals
13 Chemistry in the Marine Environment
14 Causes and Environmental Implications of Increased UV-B Radiation
15 Food Safety and Food Quality
16 Assessment and Reclamation of Contaminated Land
17 Global Environmental Change
18 Environmental and Health Impact of Solid Waste Management Activities
19 Sustainability and Environmental Impact of Renewable Energy Sources
20 Transport and the Environment

How to obtain future titles on publication

A subscription is available for this series. This will bring delivery of each new volume immediately upon publication and also provide you with online access to each title *via* the Internet. For further information visit www.rsc.org/issues or write to:

Sales and Customer Care Department, Royal Society of Chemistry, Thomas Graham House, Science Park, Milton Road, Cambridge CB4 0WF, UK

Registered Charity Number 207890

Telephone: +44 (0) 1223 432360
Fax: +44 (0) 1223 426017
Email: sales@rsc.org

ISSUES IN ENVIRONMENTAL SCIENCE AND TECHNOLOGY

EDITORS: R. E. HESTER AND R. M. HARRISON

20
Transport and the Environment

RS•C
advancing the chemical sciences

ISBN 0-85404-295-4
ISSN 1350-7583

A catalogue record for this book is available from the British Library

Published by The Royal Society of Chemistry, Thomas Graham House, Science Park, Milton Road, Cambridge CB4 0WF, UK

Registered Charity Number 207890

For further information see our web site at www.rsc.org

Typeset in Great Britain by Vision Typesetting Ltd, Manchester, UK
Printed and bound by Athenaeum Press Ltd, Gateshead, Tyne and Wear, UK

Preface

Mass transportation has become central to the lifestyle of developed societies. Long-distance commuting to work has become commonplace, and commerce requires ever greater amounts of transport capacity for goods over ever increasing distances. Alongside these trends, the advent of low-cost air travel has led to a massive increase in passenger mileage. Inevitably, such developments have consequences for the environment and, ultimately, also for human health.

Air transport is one of the fastest growing transport sectors and subsonic aircraft operate, typically, in the upper troposphere where their emissions can have an appreciable impact. David Lee reviews the contribution of aircraft to global pollution and examines some aircraft-specific phenomena such as contrail formation that have a direct impact on climate forcing. In the second chapter, Dick Derwent takes a forward look at the possible consequences of a future hydrogen economy, which is seen by some experts as a panacea largely without adverse environmental consequences. This chapter takes a dispassionate look at the effects of increasing the use of hydrogen as a fuel upon atmospheric chemistry and how this will impact upon climate. The overall conclusion, however, is that the climatic effects of a hydrogen economy will be much smaller than those of the carbon economy it may replace and, at least in respect of climate impacts, this is rather reassuring.

An ultimate aim of policy should be to render mass transport sustainable. A valuable tool in assessing progress towards sustainability is the performance indicator, which can be used as a measure of progress. However, devising such indicators is by no means straightforward and in the third chapter Henrik Gudmundsson explains the complexities of devising performance indicators for sustainable transport and reviews some of the more important activities to date. Perhaps even more problematic is the development of policy instruments for achieving sustainable transport. Most members of society have a strong belief in reducing car use, but a much lesser enthusiasm for reducing their use of their own car. Inevitably, reductions in car use will only be effected as a result of the introduction of policy instruments such as the congestion charging introduced recently in central London. In the fourth chapter David Begg, who is well known as a transport adviser to the UK government, and David Gray discuss the policy instruments that could be applied and their likely effectiveness.

Transport, especially in the form of road vehicles, contaminates not only the atmosphere but also surface waters. In the next chapter, Mike Revitt examines how traffic is responsible for a wide range of pollutants that subsequently enter and cause deterioration of the quality of surface waters. This is an area where a good deal of specific case study information is available. Engineering controls can be developed but only at considerable expense.

Evaluation of climate change as a result of changes to radiative forcing brought about by greenhouse gases and other atmospheric constituents can currently only be evaluated through the use of computationally intensive global circulation models. One of the major centres for research on climate change is the Max Planck Institute (MPI) for Meteorology in Hamburg, and Martin Schultz, Johann Feichter, and Jacques Leonardi of the MPI have contributed a chapter on climate impacts of surface transport, in which they evaluate the contribution of surface forms of transport to greenhouse emissions and the consequent effects on the atmosphere.

Whilst greenhouse gases have little direct effect on human health, there is ample evidence that other pollutants generated by combustion can have very significant effects on the health of human populations. In the final chapter, Roy Harrison and Stephen Thomas examine the contribution of the various surface transport options to emissions of locally acting air pollutants and review the evidence that those pollutants are having an impact upon public health. Specific studies of the health effects of living in close proximity to major roads are also considered.

This volume of *Issues* represents an important collection of papers on the major aspects of the subject from authors with international reputations for their research in the field. As such, it presents an authoritative review of the current state of knowledge that should prove of lasting value to scientists, policymakers and students on environmental science and engineering courses.

Roy M. Harrison
Ronald E. Hester

Contents

The Impact of Aviation on Climate **1**
David S. Lee

1 Introduction 1
2 Aircraft Emissions 4
3 Radiative Forcing and Climate Change 9
4 Aviation's Impact on Radiative Forcing and Climate Change 11
5 Reducing the Impacts 17
6 Conclusions 22
7 Acknowledgements 23

Global Warming Consequences of a Future Hydrogen Economy **25**
Richard Derwent

1 Introduction 25
2 Fate and Behaviour of Hydrogen in the Atmosphere 26
3 Hydrogen as a Greenhouse Gas 29
4 Greenhouse Gas Consequences of a Global Hydrogen Economy 32
5 Conclusions 22
6 Acknowledgements 33

Sustainable Transport and Performance Indicators **35**
Henrik Gudmundsson

1 Introduction 35
2 Making Sustainable Transport Operational 38
3 Sustainable Transport Indicator Sets 52
4 Discussion and Conclusion 61

Issues in Environmental Science and Technology, No. 20
Transport and the Environment

Policy Instruments for Achieving Sustainable Transport **65**
David Begg and David Gray

1 Introduction 65
2 'Sustainable' Transport in the UK: the Transport White Paper 66
3 From Integration to Political Pragmatism: the Evolution of Transport Policy 68
4 National Transport Policy 71
5 Local Transport Delivery 74
6 The Case for National Congestion Charging 76
7 Delivering National Congestion Charging 77
8 Conclusion 80

Water Pollution Impacts of Transport **81**
D. Mike Revitt

1 Introduction 81
2 Sources of Transport-derived Pollutants 82
3 Impacts of Transport-derived Pollutants 89
4 Control of Transport-derived Pollutants 97
5 Conclusions 108

Climatic Impact of Surface Transport **111**
Martin G. Schultz, Johann Feichter and Jaques Leonardi

1 Historical Evolution of Surface Transport 111
2 Mechanisms of Climate Impact 116
3 Emissions from Surface Traffic 120
4 Climatic Impacts of Traffic Emissions 121
5 Future Developments and Possible Mitigation Options 124
6 Conclusions 126

Human Health Implications of Air Pollution Emissions from Transport **129**
Stephen B. Thomas and Roy M. Harrison

1 Sources and Emission Inventories of the Transport Sector 129
2 Estimates of Ground-level Concentrations of Air Pollutants Attributable to Transport Emissions 134
3 Health Effects of Transport-related Air Pollutants 141
4 Methods of Quantification of the Effects of Air Pollution on Health 144
5 Recent Investigations of the Impacts of Transport Emissions upon Human Health 150

Subject Index **157**

Editors

Ronald E. Hester, BSc, DSc(London), PhD(Cornell), FRSC, CChem

Ronald E. Hester is now Emeritus Professor of Chemistry in the University of York. He was for short periods a research fellow in Cambridge and an assistant professor at Cornell before being appointed to a lectureship in chemistry in York in 1965. He was a full professor in York from 1983 to 2001. His more than 300 publications are mainly in the area of vibrational spectroscopy, latterly focusing on time-resolved studies of photoreaction intermediates and on biomolecular systems in solution. He is active in environmental chemistry and is a founder member and former chairman of the Environment Group of the Royal Society of Chemistry and editor of 'Industry and the Environment in Perspective' (RSC, 1983) and 'Understanding Our Environment' (RSC, 1986). As a member of the Council of the UK Science and Engineering Research Council and several of its sub-committees, panels and boards, he has been heavily involved in national science policy and administration. He was, from 1991 to 1993, a member of the UK Department of the Environment Advisory Committee on Hazardous Substances and from 1995 to 2000 was a member of the Publications and Information Board of the Royal Society of Chemistry.

Roy M. Harrison, BSc, PhD, DSc(Birmingham), FRSC, CChem, FRMetS, Hon MFPH, Hon FFOM

Roy M. Harrison is Queen Elizabeth II Birmingham Centenary Professor of Environmental Health in the University of Birmingham. He was previously Lecturer in Environmental Sciences at the University of Lancaster and Reader and Director of the Institute of Aerosol Science at the University of Essex. His more than 300 publications are mainly in the field of environmental chemistry, although his current work includes studies of human health impacts of atmospheric pollutants as well as research into the chemistry of pollution phenomena. He is a past Chairman of the Environment Group of the Royal Society of Chemistry for whom he has edited 'Pollution: Causes, Effects and Control' (RSC, 1983; Fourth Edition, 2001) and 'Understanding our Environment: An Introduction to Environmental Chemistry and Pollution' (RSC, Third Edition, 1999). He has a close interest in scientific and policy aspects of air pollution, having been Chairman of the Department of Environment Quality of Urban Air Review Group and the DETR Atmospheric Particles Expert Group as well as a member of the Department of Health Committee on the Medical Effects of Air Pollutants. He is currently a member of the DEFRA Air Quality Expert Group, the DEFRA Advisory Committee on Hazardous Substances and the DEFRA Expert Panel on Air Quality Standards.

Contributors

David Begg, *The Centre for Transport Policy, The Robert Gordon University, Schoolhill, Aberdeen, AB10 1RF*
Richard Derwent, *18 Kingsland Grange, Newbury, Berkshire, RG14 6LH*
Johann Feichter, *Max Planck Institute for Meteorology, Bundestr. 55, D-20146 Hamburg, Germany*
David Gray, *The Centre for Transport Policy, The Robert Gordon University, Schoolhill, Aberdeen, AB10 1RF*
Henrik Gudmundsson, *Roskilde University, Department of Environment, Technology and Social Studies, Building 10.1, Postbox 260, DK-4000, Roskilde, Denmark*
David S. Lee, *Department of Environmental and Geographical Sciences, Manchester Metropolitan University, John Dalton Extension, Chester Street, Manchester M1 5GD, United Kingdom*
Jaques Leonardi, *Max Planck Institute for Meteorology, Bundestr. 55, D-20146 Hamburg, Germany*
D. Mike Revitt, *School of Health and Social Sciences, Middlesex, University, Bounds Green Road, London N11 2NQ*
Martin G. Schultz, *Max Planck Institute for Meteorology, Bundestr. 55, D-20146 Hamburg, Germany*
Stephen B. Thomas, *Division of Environmental Health & Risk Management, School of Geography, Earth & Environmental Sciences, The University of Birmingham, Edgbaston, Birmingham B15 2TT*

The Impact of Aviation on Climate

DAVID S. LEE

1 Introduction

The atmospheric impact of aviation falls into two distinct categories: those upon the global atmosphere and those upon local air quality. Further, the impact upon the global atmosphere can be subdivided between climate change and stratospheric ozone (O_3) depletion. The latter has only been studied from a hypothetical point of view, since this is a potential effect that might arise from a fleet of supersonic aircraft flying in the mid stratosphere. This chapter will focus on the effects of aviation on the global atmosphere and, in particular, climate change. Whilst the scientific understanding of aviation's impacts on air quality is reasonably well understood, specific details that allow robust assessments of aviation impacts on local air quality — perhaps surprisingly — are only poorly quantified and the reader is directed elsewhere for a brief overview.[1]

Interest in aviation's effects on climate has been provoked by the strong growth of the aviation industry, which has outstripped GDP, long-term growth rates of the order 5% per year being sustained. Particular events have been associated with set-backs to this growth: the Gulf conflict in the early 1990s slowed growth but it picked up quickly and returned to the long-term trend within a few years. More recently, the overall economic downturn of the industry, September 11th, and 'SARS' have taken their toll but there are indications that growth rates are recovering. The seasonal patterns of departures between 1997 and 2003 in Europe (Figure 1) provide evidence of this.

Many forecasts of aviation growth have been made, both by the industry and others; typical is that of the UK Department of Trade and Industry (Figure 2), which shows historical and future projected growth to 2020 in terms of the overall capacity.

In developing this overview, and what current research is telling us, it is worth considering some historical aspects — the origins of interest date back perhaps

[1] H. L. Rogers, D. S. Lee., D. W. Raper, P. M. de Forster, C. W. Wilson and P. J. Newton, *Aeronaut. J.*, 2002, **106**, 521.

Issues in Environmental Science and Technology, No. 20
Transport and the Environment

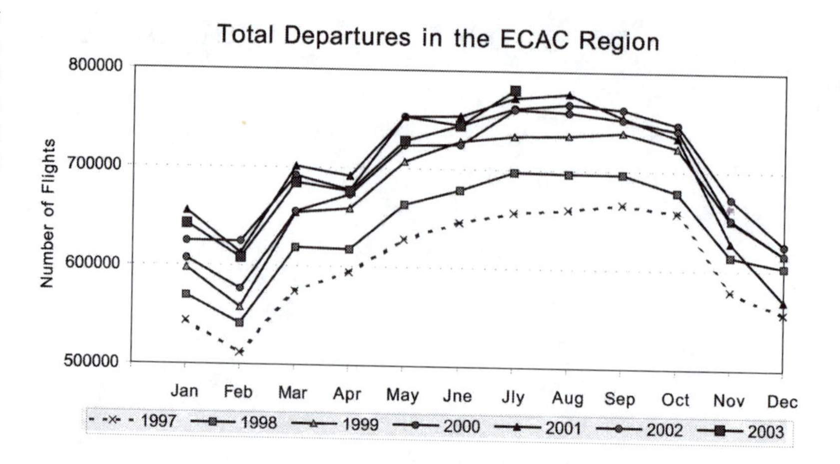

Figure 1 Monthly departures within the Eurocontrol air traffic region, 1997–2003 (data Eurocontrol, personal communication)

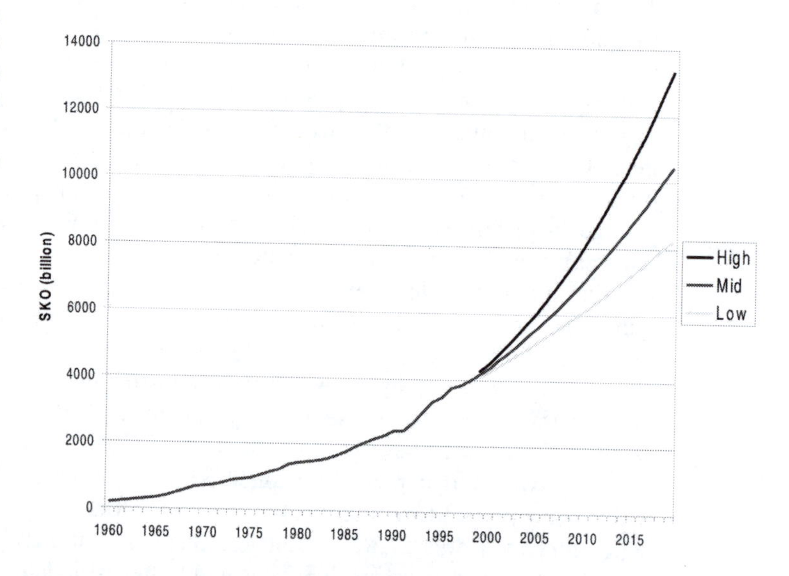

Figure 2 Aviation growth in terms of global SKO (seat kilometres offered) between 1960 and 2020 (source: DTI data, personal communication)

further than one might suspect. Local air quality was the original driver for the development of aircraft engine emissions regulations by the International Civil Aviation Organization (ICAO), first promulgated in 1981 (although earlier local rules to the USA were introduced by the US Environmental Protection Agency in 1973). However, one of the initial drivers of interest in aviation's atmospheric effects was concern in the late 1960s and early 1970s that emissions of nitrogen oxides (NO_x, where $NO_x = NO + NO_2$) from a (proposed) fleet of supersonic aircraft flying in the stratosphere would significantly deplete stratospheric ozone (O_3), resulting in increased exposure of harmful ultraviolet (UV) radiation at the Earth's surface.[2,3] The scientific research programmes that this concern initiated

[2] H. S. Johnston, *Science*, 1971, **173**, 517.
[3] P. J. Crutzen, *Ambio*, 1972, **1**, 41.

were of quite epic proportions and laid many of the modern foundations of our understanding of stratospheric chemistry and physics. A more detailed account of these research programmes, and their development, is given elsewhere.[4] In fact, the US research programme proceeded *after* the decision had been taken in the US not to build a supersonic transport (or 'SST') and was partly in response to the intentions of the UK and France to build Concorde, and the USSR the Tupolev TU-144. During this early work it was conjectured that the current subsonic fleet may, in fact, impact upon tropospheric O_3, following the proposal of Crutzen[5] that *in situ* production dominated tropospheric O_3.

Interest in the potential effects of subsonic aviation ensued in the 1980s and early 1990s.[6] This interest arose because of the growing realization that the upper troposphere and lower stratosphere, where subsonic aircraft cruise, is a rather sensitive region of the atmosphere in terms of its chemistry. Initially, attention was focussed upon the effects of aircraft NO_x emissions on tropospheric O_3 production. Whereas O_3 in the mid to upper stratosphere provides a protective 'shield' against harmful UV radiation, O_3 in the upper troposphere and lower stratosphere acts as a powerful greenhouse gas, warming the Earth's surface. More recently, other effects such as those of contrails (condensation trails) have been studied intensively, although studies of contrails and climate can be traced back to the early 1970s.[7]

Contrails are line-shaped ice clouds caused by the emission of water vapour and particles from the aircraft exhaust. Depending principally on the particular conditions of temperature and humidity (strictly, ice-supersaturation), contrails may be very short-lived or persistent, sometimes spreading by wind-shear, sedimentation and diffusion into cirrus-like clouds that are ultimately unrecognizable as having been caused by aircraft. Other effects on climate from associated particle emissions and the enhancement of cirrus clouds have also been suggested.

In 1996, the Intergovernmental Panel on Climate Change (IPCC), at the request of ICAO, announced its intention to assess aviation's effects on the global atmosphere; this was completed in 1999.[8] However, the IPCC was not the first assessment: other previous assessments and syntheses include, *e.g.* Schumann (1994),[9] Wahner *et al.* (1995),[10] Friedl *et al.* (1997),[11] Brasseur *et al.* (1998).[12] This

[4] D. S. Lee, *Annex 1* to *QinetiQ report QINETIQ/FST/CR030440*, D. H. Lister and P. D. Norman (eds.), 2003.

[5] P. J. Crutzen, *Tellus*, 1974, **26**, 47.

[6] U. Schumann, *Air Traffic and the Environment*, Lecture Notes in Engineering, No. 60, ed. U. Schumann (ed.), Springer-Verlag, Berlin, 1990.

[7] D. R. Lyzenga, PhD thesis, University of Michegan, Ann Arbor, 1973.

[8] IPCC, *Aviation and the Global Atmosphere.* A Special Report of IPCC Working Groups I and III in collaboration with the Scientific Assessment panel to the Montreal Protocol on Substances that Deplete the Ozone Layer, J. E. Penner, D. H. Lister, D. J. Griggs, D. J. Dokken and M. McFarland (eds.), Intergovernmental Panel on Climate Change, Cambridge University Press, Cambridge, 1999.

[9] U. Schumann, *Ann. Geophys.*, 1994, **12**, 365.

[10] A. Wahner, M. A. Geller, F. Arnold, W. H. Brune, D. A. Cariolle, A. R. Douglass, C. Johnson, D. H. Lister, J. A. Pyle, R. Ramaroson, D. Rind, F. Rohrer, U. Schumann and A. M. Thompson, Subsonic and supersonic aircraft emissions, In *Scientific Assessment of Ozone Depletion: 1994* World Meteorological Organization Global Ozone Research and Monitoring project—Report No. 37, Geneva, 1995.

period saw tremendous activity originating from national/international research programmes and dedicated efforts for the IPCC report. Shortly before the completion of the IPCC assessment, Boeing announced that it no longer intended to pursue the development of an SST, largely on economic and environmental (noise) grounds. This, along with the overspending and overrunning NASA space station programme, precipitated the termination of NASA's Atmospheric Effects of Aviation Programme (AEAP). Subsequently, some activities were restarted in the US, albeit at a much lower budgetary level, primarily on global modelling and engine emissions. In Europe, however, the IPCC aviation report[8] provided a springboard from which several research programmes into atmospheric science and technology were initiated under the European Commission's Fifth Framework Programme and included: PARTEMIS, NEPAIR, TRADEOFF, INCA, AERO2K, SCENIC and CRYOPLANE.* The bulk of the efforts of these programmes were directed at subsonic effects/technology, with the exception of SCENIC and minor components of TRADEOFF, which addressed supersonic impacts.

In Section 2, the emissions from aircraft in terms of species and their global nature are described. Section 3 gives a brief description of the climate metric, radiative forcing, followed by specific aviation impact quantification (Section 4). In Section 5, some potential emissions reduction approaches are described and some brief conclusions drawn in Section 6.

2 Aircraft Emissions

Aircraft Engine Emissions

The civil subsonic fleet is dominated by aircraft equipped with turbofan gas turbine engines; the turboprop fleet being relatively small on a global scale. Gas turbine engines are technologically advanced systems that require stringent characteristics of safety and durability. Engine emissions are regulated through certification requirements of ICAO (ICAO, 1981;[13] most recent update ICAO, 1995[14]) for NO_x, unburned hydrocarbons (HCs), carbon monoxide (CO), and smoke. It is worth reinforcing what these regulations are: they are manufacturing standards, not an in-service compliance regime. Thus, measurements are made using carefully prescribed methodologies on a limited number of engines for certification purposes.

In recent years, significant improvements in fuel efficiency have been achieved

* See *e.g.*, http://www.ozone-sec.ch.cam.ac.uk/clusters/Corsaire_Website/corsaire_index.htm

[11] R. R. Friedl, S. L. Baughcum, B. Anderson, J. Hallett, K.-N. Liou, P. Rasch, D. Rind, K. Sassen, H. Singh, L. Williams and D. Wuebbles, *Atmospheric Effects of Subsonic Aircraft: Interim Assessment of the Advanced Subsonic Assessment Program*, NASA Reference Publication 1400, Washington D.C., 1997.

[12] G. P. Brasseur, R. A. Cox, D. Hauglustaine, I. Isaksen, J. Lelieveld, D. H. Lister, R. Sausen, U. Schumann, A. Wahner and P. Wiesen, *Atmos. Environ.*, 1998, **32**, 2329.

[13] ICAO, *International Standards and Recommended Practices, Environmental Protection, Annex 16 to the Convention on International Civil Aviation, Vol. II, Aircraft Engine Emissions*, (1st edn.), International Civil Aviation Organization, Montreal, 1981.

[14] ICAO, *ICAO Engine Exhaust Emissions Databank*, 1st Edn., ICAO Doc. 9646-AN/943, International Civil Aviation Organization, Montreal, 1995.

to reduce operating costs. Also, emissions of some pollutants, particularly smoke, have been reduced. Emissions of CO_2 and H_2O scale with fuel consumption depending on the specific fuel carbon to hydrogen ratio. Emissions of NO_x and soot are highest at high power settings whilst CO and HCs are highest at low power settings as they are the result of incomplete combustion. In general, emissions of NO_x, CO, HCs and particles are relevant to local air quality issues and CO_2, H_2O, NO_x, SO_x and particles are of most concern in terms of climate perturbation. The production and control of these emissions are described briefly below. For a more detailed account, the reader is referred to other reviews.[8,12]

Oxides of Nitrogen (NO_x). Emissions of NO_x arise from the oxidation of atmospheric nitrogen in the high temperature conditions that exist in the engine's combustor, although a small amount comes from the nitrogen content of the fuel. Its production is a complex function of combustion temperature, pressure and combustor design. Although NO_x emissions can be reduced, as overall engine pressure ratios have increased (to reduce fuel consumption), this has implied higher temperatures and pressures in the combustor, which tend to increase NO_x production. Hence, to address NO_x emissions, different combustor technologies have been developed.[8] Nitrogen oxides at the engine exit plane consist primarily of NO. The percentage of NO_2 to NO is estimated to be 1–10%, with an uncertainty of several percent. However, NO is quickly converted into NO_2 in the atmosphere.

Particles. Particles emitted from aircraft can be categorized into volatile and non-volatile components; this being partially an operational measurement definition. Non-volatile particles primarily include carbonaceous material formed in the primary combustion zone arising from incomplete combustion of the fuel. A fleet average emission index (EI) for soot of 0.04 g per kg fuel burned has been estimated[15] with a large uncertainty (at best a factor of 2). These soot particles are thought to have only a minor direct impact upon climate (see the Section on Sulfate and Soot Particles). However, soot and other particles emitted from aircraft engines play a role in contrail and cirrus cloud enhancement, as shown later. Soot particles *per se* are not regulated but rather the so-called 'Smoke Number',[13,14] which is an optical measurement of, effectively, the larger soot particles.

Volatile particles are primarily composed of sulfate, although recent research suggests that some smaller fraction of these particles is composed of organic material.[16] Most of the sulfur in the fuel is expected to be emitted as sulfur dioxide (SO_2).[17] However, some oxidation through to S^{VI} (*e.g.* SO_3, H_2SO_4) is possible within the engine itself.[18] The fraction of total gaseous sulfur in the engine exit plane is estimated to be up to 5% S^{VI}: however, this estimate is highly

[15] A. Döpelheuer, SAE Paper No. 2001-01-3008, Proceedings of the 2001 Aerospace Congress, September 10–14, 2001.
[16] B. Kärcher, *Atmos. Res.*, 1998, **46**, 293.
[17] R. C. Miake-Lye, M. Martinez-Sanchez, R. C. Brown and C. E. Kolb, *J. Aircraft*, 1993, **30**, 467.
[18] S. P. Lukachko, I. A. Waitz, R. C. Miake-Lye, R. C. Brown and M. R. Anderson, *J. Geophys. Res.*, 1998, **103**, 16159.

Table 1 Fuel, CO_2, NO_x and $EINO_x$ for 1991/92, 1999, 2000, 2015 and 2050 gridded data sets

Dataset	*Fuel ($Tg\ yr^{-1}$)*	*CO_2 ($Tg\ C\ yr^{-1}$)*	*NO_x (as NO_2) ($Tg\ yr^{-1}$)*	*$EINO_x$*	*Reference*
ANCAT/EC2 – 1991/92	131.3	113	1.81	13.8	Gardner *et al.*, 1998[32]
TRADEOFF 2000				13.8	TRADEOFF (2003)[46]
ANCAT/EC2 – 2015	286.9	247	3.53	12.3	Gardner *et al.*, 1998[32]
NASA 1992				12.6	Baughcum *et al.*, 1996[34]
NASA 1999				13.2	Sutkus *et al.*, 2001[35]
NASA – 2015				13.7	Baughcum *et al.*, 1998[37]
FESG Fa1 – 2050	471.0	405	7.2	15.2	FESG, 1998[38]
FESG Fa2 – 2050	487.6	419	5.5	11.4	FESG, 1998[38]
FESG Fc1 – 2050	268.2	231	4.0	15.0	FESG, 1998[38]
FESG Fc2 – 2050	277.2	238	3.1	11.3	FESG, 1998[38]
FESG Fe1 – 2050	744.3	640	11.4	15.3	FESG, 1998[38]
FESG Fe2 – 2050	772.1	664	8.8	11.4	FESG, 1998[38]

uncertain. Emission of S species is thought to be important for volatile particle formation from sulfuric acid (H_2SO_4) and gas-phase H_2SO_4 has now been detected in the wake of aircraft.[19]

Measurements of particle emissions from aircraft, engines and combustors have shown that they lie in the 3 nm to 4 μm aerodynamic diameter size range. The soot aerosol size distribution at the engine exit is log–normal, with number concentrations peaking in the 40–60 nm size range. Emission indices fall within the range of 10^{12} soot aerosol particles per kg fuel for current advanced combustors and up to 10^{15} for older engines.[20]

Other Trace Species. Other trace species have not been as well characterized as, for example, NO_x. Hydroxyl radicals (OH) are produced as a part of the combustion process and control the oxidation of NO_x and S species to their oxidized forms. Few measurements of OH have been made, despite its importance.[21] Tremmel *et al.* (1998)[22] used measurements of other odd N species in the plume to infer OH concentrations of 1 ppmv or less. However, recent static measurements using Laser Induced Fluorescence (LIF) indicated concentrations of 100 ppbv or less.[23] Recent measurements from the PARTEMIS study indicate much lower OH concentrations, of the order 1 ppb.[24] Clearly, given the importance of OH,

[19] J. Curtuis, B. Sierau, F. Arnold, R. Baumann, R. Busen, P. Schulte and U. Schumann, *Geophys. Res. Lett.*, 1998, **25**, 923.

[20] A. Petzold, A. Döpelheuer, C. A. Brock and F. P. Schröder, *J. Geophys. Res.*, 1999, **104**, 22171.

[21] T. F. Hanisco, P. O. Wennberg, R. C. Cohen, J. G. Anderson, D. W. Fahey, E. R. Keim, R. S. Gao, R. C. Wamsley, S. G., Donnelly, L. A. DelNegro, R. J. Salawitch, K. K. Kelly and M. H. Proffitt, *Geophys. Res. Lett.*, 1997, **24**, 65.

[22] H. G. Tremmel, H. Schlager, P. Konopka, P. Schulte, F. Arnold, M. Klemm and B. Droste-Franke, *J. Geophys. Res.*,1998, **103**, 10803.

[23] S. Bockle, S. Einecke, F. Hildenbrand, C. Orlemann, C. Schulz, J. Wolfrum and V. Sick, *Geophys. Res. Lett.*, 1999, **26**, 1849.

[24] C. W. Wilson, A. Petzold, S. Nyeki, U. Schumann and R. Zellner, *Aerosp. Sci. Technol.*, 2004, **8**(2), 131.

more measurements are needed to provide upper limits for oxidation of various species within the engine and the plume.

Emissions of nitrous (HONO) and nitric (HNO_3) acids relative to total NO_y are estimated to be less than a few percent[25] but emission indices are rather uncertain. More measurements are needed to better quantify this speciation.

The high temperatures involved in combustion of kerosene within aircraft gas turbines produces gaseous ions by chemiionisation of free radicals, often termed 'chemi-ions' (CIs). The first measurements of negative ions ($HSO_4^-H_2SO_4$, $HSO_4^-HNO_3$) were made behind an aircraft engine at ground level.[26] Subsequently, other measurements have shown the presence of negative CIs in the plume of an aircraft exhaust,[27] and also positive CIs.[28] CIs may promote formation and growth of charged droplets.[29]

Global Aircraft Emissions Characterization

Several estimations of global aviation emissions have been made over the past ten years or so have been summarized in the IPCC report.[8,30] A more recent discussion has been provided by Lee *et al.* (2002).[31] Global inventories of aircraft emissions usually provide 3D gridded data and, by necessity, have simplifying assumptions. The essential components of an inventory include: an aircraft movements database; a representation of the global fleet in terms of aircraft and engines; a fuel-flow model; a method for calculation of emissions; and landing and take-off emissions data.

The emissions data in most common usage were calculated for the early 1990s, *e.g.* the ANCAT/EC2 dataset;[32] the DLR-2 data set;[33] and the NASA data set.[34]

[25] F. Arnold, J. Scheid, T. Stilp, H. Schlager and M. E. Reinhardt, *Geophys. Res. Lett.*, 1992, **19**, 2421.

[26] A. Frenzel and F. Arnold, Sulphur acid cluster ion formation by jet engines: implications for sulfuric acid formation and nucleation, in *Impact of Emissions from Aircraft and Spacecraft on the Atmosphere*, Proceedings of an International Scientific Colloquium, Köln (Cologne), Germany, April 18-20, 1994, U. Schumann and D. Wurzel (eds.), DLR Mitteilung 94-06, Köln, 1994.

[27] F. Arnold, J. Curtuis, B. Sierau, V. Bürger, R. Busen and U. Schumann, *Geophys. Res. Lett.*, 1999, **26**, 1577.

[28] K. H. Wohlfrom, S. Eichkorn, F. Arnold and P. Schulte, *Geophys. Res. Lett.*, 2000, **27**, 3853.

[29] F. Q. Yu and R. P. Turco, *Geophys. Res. Lett.*, 1997, **24**, 1927.

[30] S. C. Henderson, U. K. Wickrama, S. L. Baughcum, J. L. Begin, F. Franco, D. L. Greene, D. S. Lee, M. L. Mclaren, A. K. Mortlock, P. J. Newton, A. Schmitt, D. J. Sutkus, A. Vedantham and D. J. Wuebbles, *Aviation and the Global Atmosphere*, J. E. Penner, D. H. Lister, D. J. Griggs, D. J. Dokken and M. McFarland (eds.), A Special Report of IPCC Working Groups I and III in collaboration with the Scientific Assessment panel to the Montreal Protocol on Substances that Deplete the Ozone Layer, Intergovernmental Panel on Climate Change, Cambridge University Press, Cambridge, 1999, Ch. 9.

[31] D. S. Lee, B. Brunner, A. Döpelheuer, R. S. Falk, R. M. Gardner, M. Lecht, M. Leech, D. H. Lister and P. J. Newton, *Meteorol. Z.*, 2002, **11**, 141.

[32] R. M. Gardner, J. K. Adams, T. Cook, L. G. Larson, R. S. Falk, E. Fleuti, W. Förtsch, M. Lecht, D. S. Lee, M. V. Leech, D. H. Lister, B. Massé, K. Morris, P. J. Newton, A. Owen, E. Parker, A. Schmitt, H. ten Have and C. Vandenberghe, *ANCAT/EC2 aircraft emissions inventories for 1991/1992 and 2015: Final Report*, Produced by the ECAC/ANCAT and EC Working Group, European Civil Aviation Conference, 1998.

[33] A. Schmitt and B. Brunner, in Final Report on the BMBF Verbundprogramm, *Schadstoffe in der Luftfahrt*, U. Schumann, A. Chlond, A. Ebel, B. Kärcher, H. Pak, H. Schlager, A. Schmitt and P. Wendling (eds.), DLR Mitteilung 97-04, Köln, 1997.

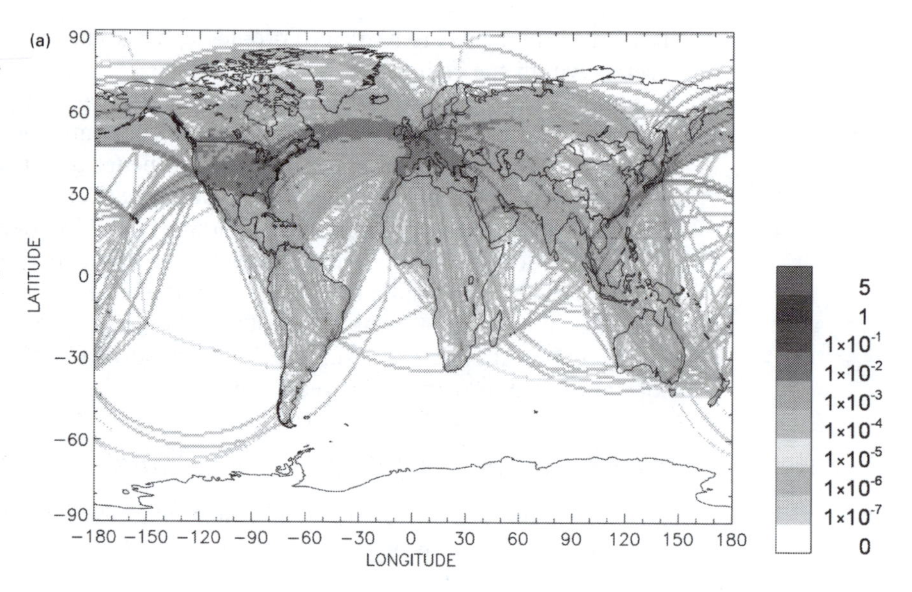

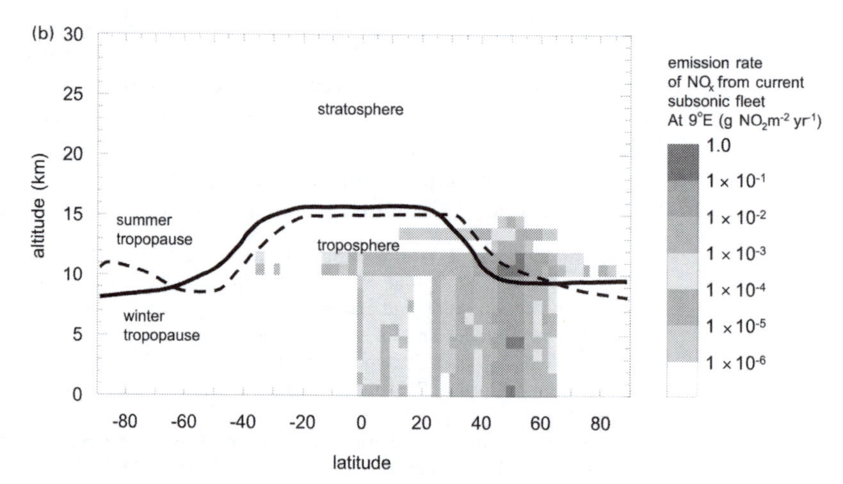

Figure 3 (a) Spatial distribution of ANCAT/EC2 1991/92 emissions of NO_x from civil aviation, vertically integrated between ground and 16 km (kg NO_2 m^{-2} yr^{-1}). Source: Gardner *et al.*[32] (b) Emission rate of the 1992 aircraft fleet from longitude 9° E in relation to the main structure of the troposphere and stratosphere

More recently, Boeing have produced an estimate for 1999[35] and within the TRADEOFF project an estimate was made for 2000 emissions by scaling up the 1992 ANCAT/EC2 traffic data according to regional traffic statistics.[36]

For future global emissions estimations, there are emission forecasts available for 2015[32,37] and scenarios for 2050.[38]

[34] S. L. Baughcum, T. G. Tritz, S. C. Henderson and D. C. Pickett, *Scheduled Civil Aircraft Emissions Inventories for 1992: Data base Development and Analysis*, NASA Contractor report 4700, NASA Langley Research Center, US, 1996.

[35] D. J. Sutkus, S. L. Baughcum and D. P. DuBois, *Scheduled Civil Aircraft Emission Inventories for 1999: Database Development and Analysis*, NASA Contractor Report 2001-211216, NASA Glenn Research Center, US, 2001.

[36] D. S. Lee, R. D. Kingdon, P. D. Norman and R. Sausen, manuscript in preparation, 2004.

[37] S. L. Baughcum, S. C. Henderson and D. J. Sutkus, *Scheduled Civil Aircraft Emission Inventories Projected for 2015: Database Development and Analysis*, NASA/CR-1998-207638, NASA Langley Research Center, US, 1998.

Estimations of global fuel and NO_x for 1991/92, 1999, 2000, 2015 and 2050 are given in Table 1. Emissions from the military fleet are much more uncertain than the civil estimates but are approximately 10% of the global total. For climate effects other than CO_2, military aviation is less important as most emissions are at much lower altitudes than those from civil aircraft.[32]

The horizontal and vertical structure of aircraft emissions is shown in Figures 3(a) and 3(b), respectively. These figures clearly depict how the overall pattern of emissions is concentrated in the Northern Hemisphere at altitudes between 8 and 12 km.

By 2015, it is forecast that emissions of NO_x (as NO_2) from civil aviation will have grown by a factor of nearly 2, to 3.53 Tg N yr^{-1} with a fuel burn of 287 Tg yr^{-1}. For 2050, the most commonly recognized data sets are those generated by the Forecasting and Economic Sub Group[38] of the ICAO for the IPCC aviation Special Report.[8] These scenarios were based upon a relationship between revenue passenger kilometres (RPK) and GDP. The GDP scenarios were the IPCC[†] 'IS92a, c and e' scenarios.[39] In addition, two technology scenarios for fuel and NO_x were assumed, one ambitious, one less so, giving six scenarios in total. The nomenclature adopted was, for example, Fa1–FESG IS92a Technology Scenario 1. Fuel usages ranged from 268 to 772 Tg yr^{-1} and NO_x emissions by 3.1 to 11.4 Tg NO_2 yr^{-1}; these estimates equate to factors of approximately 2 to 6 on fuel and NO_x emissions over early 1990s data. By 2050, assumptions in GDP growth are clearly critical to the overall emission, the technology assumptions having a second-order effect.

Updating these inventories is a major task, as revised global traffic data need to be incorporated. Moreover, the fuel and emissions factors used need to be calculated independently, either from manufacturer's data — which are usually proprietary — or from accepted engine models. There are a number of recent initiatives that will provide updated data in 2003–2004.

3 Radiative Forcing and Climate Change

Careful analyses of palaeoclimatological data have shown that climate has changed over long time-scales. However, more recent data (both direct and proxy) have revealed a remarkable rise in global average surface temperature since the industrial revolution. That such an increase exists is beyond doubt — nonetheless attribution of this temperature increase to greenhouse gas emissions remains a contentious issue and a challenging scientific problem. To assess climate change, the World Meteorological Organization and the United Nations Environment Programme jointly established the IPCC to provide authoritative international assessments of climate change. The IPCC published

[†] IPCC IS92 scenarios have subsequently been replaced by the 'SRES' scenarios, see IPCC (2000)[43]

[38] FESG, *Report 4. Report of the Forecasting and Economic Analysis Sub-Group (FESG): Long-range scenarios*, International Civil Aviation Organization Committee on Aviation Environmental Protection Steering Group Meeting, Canberra, Australia, January 1998.

[39] J. Leggett, W. J. Pepper and R. J. Swart, In *Climate Change 1992. The Supplementary Report to the IPCC Scientific Assessment*, J. T. Houghton, B.A. Callander and S.K. Varney (eds.), Cambridge University Press, Cambridge, 1992.

major reports on the science of climate change in 1990,[40] 1995[41] and 2001.[42] In addition, IPCC has also assessed impacts and mitigation.

In the present context, the United Nations Framework Convention on Climate Change (UNFCCC) definition of climate change is used, *i.e.* a change of climate that is attributed directly or indirectly to human activity that alters the composition of the global atmosphere and that is in addition to natural climate variability observed over comparable time periods. This change may result from emissions of a pollutant from fuel combustion or, *e.g.*, as a result of land use change that results in changes in 'natural' emissions. Several factors have perturbed climate since the industrial revolution, both natural and man-induced. Separating these factors and quantifying them is difficult, as climate has a natural variability and our knowledge of these factors in not necessarily complete. The IPCC concluded in its *Summary for Policymakers* of its Third Assessment Report of Working Group 1:[42]

'*An increasing body of observations gives a collective picture of a warming world and other changes in the climate system*'.

The IPCC Third Assessment Report[42] concluded that the global average surface temperature over the 20th Century has increased by 0.6 ± 0.2°C. Other manifestations of climate change have been observed and quantified, *e.g.* decreases in snow and ice cover extent, sea level rise, and ocean heat content.
Computer models (Global Climate Models, or GCMs) are used to simulate climate and its change: separation of the various factors that control climate is extremely difficult because of representation of particular processes (*e.g.* clouds) and a number of problems including of internal variability within the models, model-to-model variability and signal-to-noise ratios. Using such GCMs and simplified derivative models, the IPCC predicted that globally averaged surface temperatures will further increase by between 1.4 to 5.8°C from 1990 to 2100, depending upon the emission scenario assumed.

As the manifestation of climate change is a complex phenomenon, it has been necessary to develop a metric that allows us to quantify it in relatively simple terms but also allows a comparison of different climate change agents. Initially, the concept of Global Warming Potentials (GWPs) was introduced;[40] an index that allowed the climate effects of emissions to be compared relative to those of CO_2. However, the limitations of GWPs became obvious for influences other than long-lived gases such as carbon dioxide (CO_2), methane (CH_4) and the halocarbons. Subsequently, the concept of *radiative forcing of climate* was commonly used as a metric of climate change by the IPCC in an interim report[44] for the first

[40] IPCC, *Climate Change the IPCC Scientific Assessment*, Report prepared for IPCC by Working Group 1, J. T. Houghton, G. J. Jenkins and J. J. Ephraums (eds.). Cambridge University Press, Cambridge, 1990.

[41] IPCC, *Climate Change 1995, the Science of Climate Change*, Contribution of Working Group I to the Second Assessment Report of the Intergovernmental Panel on Climate Change, J. T. Houghton, L. G. Meira Filho, B. A. Callander, E. Haites, N. Harris, A. Kattenberg and K. Maskell (eds.), Cambridge University Press, Cambridge, 1996.

[42] IPCC, *Climate Change 2001, the Scientific Basis, Summary for Policymakers and Technical Summary of the Working Group I Report*, Cambridge University Press, Cambridge, 2001.

[43] IPCC, *Emission Scenarios, A Special Report of Working Group III of the Intergovernmental Panel on Climate Change*, Cambridge University Press, Cambridge, 2000.

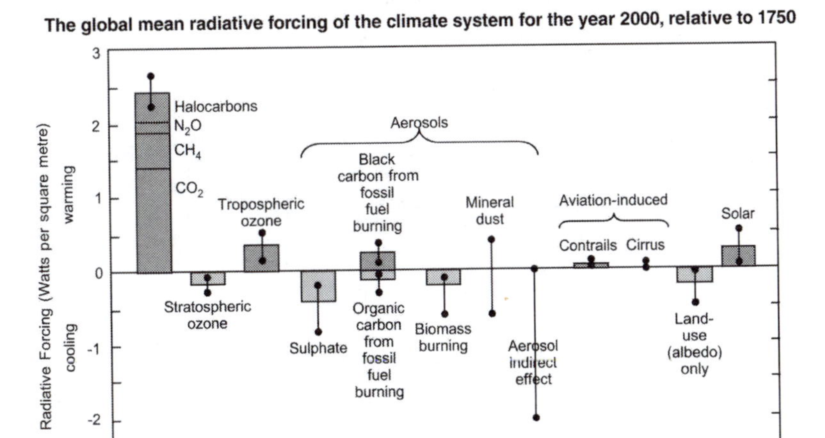

Figure 4 Global annual mean radiative forcings ($W\ m^{-2}$) from a number of agents for the period from the pre-industrial (1750) to the present (2000). Source: IPCC (2001)[42]

meeting of the Conference of Parties to the UNFCCC and the Second Assessment Report of Working Group 1 of the IPCC.[41]

Radiative forcing may be defined as a measure of the importance of perturbations to the planetary radiation balance and is measured in watts per square metre. A main reason for its use is that there is an approximately linear relationship between the change in global mean radiative forcing (ΔF) and the global mean surface temperature change (ΔT_s):

$$\Delta T_s \approx \lambda \Delta F \tag{1}$$

where λ is the climate sensitivity parameter $[K\ (W\ m^{-2})^{-1}]$.[45] Positive radiative forcing values imply warming, and negative, cooling. This is a convenient metric but nonetheless remains a proxy for climate change. The recent IPCC Third Assessment Report[42] provided the most up-to-date assessment of radiative forcing of climate from several different agents (Figure 4). Although radiative forcings are not necessarily additive, the forcing from the long-lived greenhouse gases in the first bar (*i.e.* CO_2, CH_4, N_2O, CFCs) is 2.43 $W\ m^{-2}$. This issue is dealt with further in Chapter 6.

4 Aviation's Impacts on Radiative Forcing and Climate Change

Having illustrated the total effect of human activities on climate and radiative forcing, we now consider the radiative forcing from subsonic aviation in 1992 and 2050, as estimated by the IPCC,[8] shown in Figure 5.

44 IPCC, *Climate Change 1994, Radiative Forcing of Climate Change and an Evaluation of the IPCC IS92 Emission Scenarios*, Reports of Working Groups I and III of the Intergovernmental Panel on Climate Change, forming part of the IPCC Special Report to the first session of the Conference of the Parties to the UN Framework Convention on Climate Change, J. T. Houghton, L. G. Meira Filho, J. Bruce, H. Lee, B. A. Callander, E. Haites, N. Harris and K. Maskell (eds.), Cambridge University Press, Cambridge, 1995.

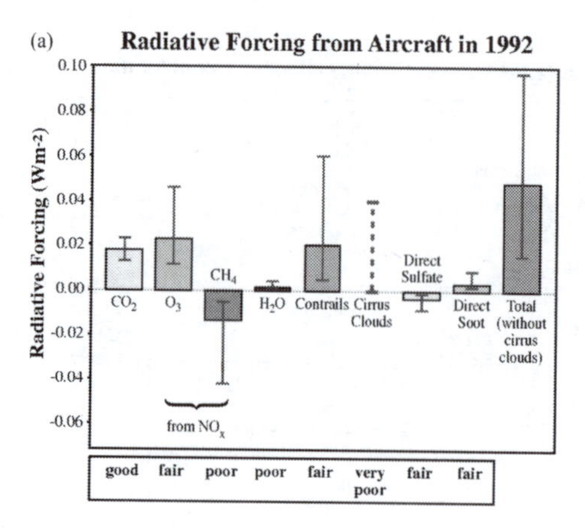

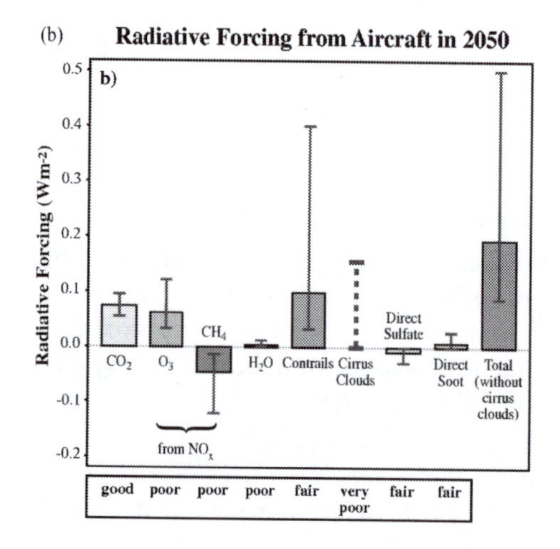

Figure 5 Global Annual Mean Radiative Forcings from Aircraft Emissions for 1992 (a) and 2050, scenario Fa1 (b) (W m^{-2}). Source: IPCC (1999)[8] and DLR web-site

Here, we see that aviation constituted an overall forcing of 0.05 W m^{-2} in 1992 (excluding any estimate from cirrus cloud enhancement), approximately 3.5% of the overall radiative forcing (as estimated in the IPCC's Second Assessment Report,[41] and an overall forcing of 0.19 W m^{-2} in 2050—5% of all climate forcing for a central scenario (IS92a). The full range of 2050 scenarios studied imply forcing ranging from 0.13 to 0.56 W m^{-2}, *i.e.* 2.6 to 11 times the value in 1992.

Figure 6 shows the large uncertainties associated with contrail and cirrus effects (the two being linked): clearly, the contrail and cirrus effects are relatively large and rather uncertain, the latter having no best estimate associated with it.

Since the seminal IPCC aviation report,[8] research has continued into the range of effects associated with aviation. A number of studies have focussed on improved estimates of various radiative forcing effects, particularly contrails. The EC TRADEOFF Project[46] has re-estimated the suite of radiative forcings from aviation, (shown in Figure 6 for 2000 and 2050).

Comparing Figures 6 and 5 shows that this recent assessment has resulted in

[45] K. P. Shine and P. M. de F Forster, *Glob. Planet. Change*, 1999, **20**, 205.

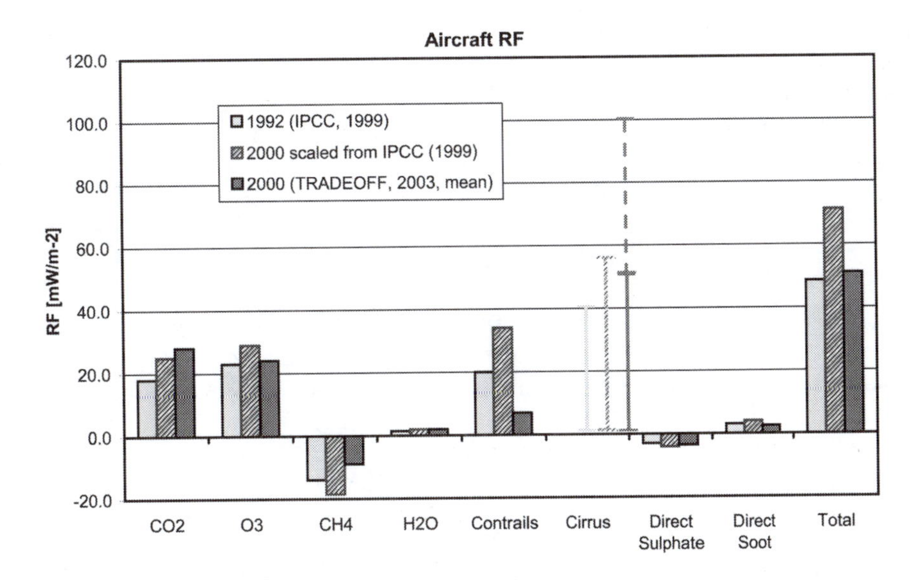

Figure 6 Radiative forcing from aviation, TRADEOFF 2000 estimates (source: EC TRADEOFF, 2003[46])

some significant changes. Below, the individual effects are considered along with the recent changes in estimates.

Carbon Dioxide

Carbon dioxide is emitted as a fixed fraction of the fuel burned, so that increases in fuel usage will result in a linear increase in CO_2 emission, which for 2050 (scenario Fa1) is projected to be a factor of 3 increase over 1992.

Since CO_2 has a long atmospheric lifetime, of the order of many decades, any emission becomes completely mixed within the earth's atmosphere. Long-term observations have shown a steady increase in atmospheric CO_2 concentrations from around 330 ppm (parts per million) in 1973 to around 370 ppm in 2000. This may be placed in the context of an increase from around 280 ppm prior to the industrial revolution. Aviation is currently estimated to produce 2% of all man-made CO_2 emissions, representing 13% of the CO_2 emission from all transportation sectors. The IPCC aviation Special Report[8] estimated that the radiative forcing from CO_2 from aviation was 0.018 W m^{-2} (range 0.013–0.023 W m^{-2}). The recent TRADEOFF estimate for 2000 gives CO_2 radiative forcing as 0.028 W m^{-2}: this is in agreement with model results shown in Figure 7, which shows the time-dependent evolution of CO_2 concentration and radiative forcing attributable to aviation from 1940 to 2050 (Fa1 scenario).

Ozone, Methane and Nitrogen Oxides

Aircraft emissions of NO_x result in enhancement of tropospheric O_3 and reductions in CH_4. This is the result of complex tropospheric chemistry: an overview of this chemistry is beyond the scope of this chapter but many reviews are available (*e.g.* References 1, 8 and 12).

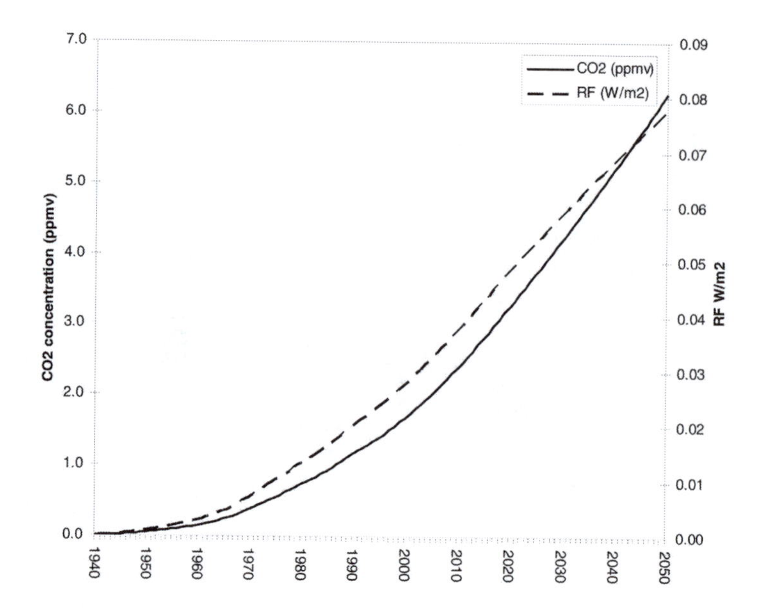

Figure 7 Evolution of CO_2 concentration and its radiative forcing from aviation, 1940–2050

The formation of tropospheric O_3 from aircraft NO_x emissions is made particularly effective by the height of emission, which results in a longer lifetime for NO_x such that, on a per mole basis, more O_3 is produced compared with surface releases. Moreover, the radiative forcing from O_3 is highly effective at the altitudes at which aircraft cruise[8]. However, aircraft emissions of NO_x result in a small enhancement of OH, which ultimately results in a small destruction of ambient CH_4. Thus, aircraft emissions of NO_x result in a positive radiative forcing from O_3 and a negative forcing from ambient CH_4 destruction. These forcings do not cancel in terms of their effect on climate since they have quite different latitudinal effects because of their different lifetimes; O_3 a few weeks to months at 200 hPa; CH_4 of the order 8–10 years.

The point about the inhomogeneous distribution of O_3 is important as there is evidence that the same radiative forcing from O_3 placed solely in the Northern Hemisphere compared with its distribution across the hemispheres produces a stronger increase in surface temperature;[47] *i.e.* a different *climate* response for the same radiative forcing. The only 'cancellation' that can be inferred from reductions in CH_4 concentrations (and consequent radiative forcing) from aircraft NO_x emissions is with CO_2, as both CO_2 and CH_4 are long-lived gases, well mixed across the hemispheres.

The TRADEOFF estimate of O_3 radiative forcing is somewhat smaller than that given by IPCC; likewise, the CH_4 forcing is also smaller. This is considered to be due to improved models with increased vertical and horizontal resolution, making them less diffusive.[46]

[46] TRADEOFF, *TRADEOFF summary report, Aircraft emissions: Contributions of various climate compounds to changes in composition and radiative forcing — tradeoffs to reduce atmospheric impact*, EU contract no. EVK2-CT-1999-0030, 2003.

Water Vapour

Although water vapour (H_2O) is a powerful greenhouse gas and is responsible for making the Earth's surface habitable, the amounts of H_2O added to the atmosphere by subsonic aircraft (current or projected in 2050) is minute, compared with the natural hydrological cycle. Thus, the H_2O emissions, *per se*, have little effect on climate. There has been no further work that suggests that the IPCC (1999)[8] estimate should be revised, thus the TRADEOFF estimate[46] is the same.

Notably, whilst additional H_2O from subsonic aviation has an insignificant effect on radiative forcing, this is not the case for supersonic scenarios in which the emissions occur in the dry stratosphere. A 2050 supersonic scenario given by the IPCC (1999)[8] showed that the total radiative forcing for 2050 of 0.19 W m^{-2} (scenario Fa1) increased to 0.27 W m^{-2}, an increase of 42%, largely from the impact of water vapour.

Contrails

Contrails (condensation trails) are ice clouds formed initially by water vapour depositing on particle emissions from aircraft, although the bulk of the water in a persistent contrail comes from the ambient atmosphere. Contrails only form in very cold and humid atmospheres (typically -35 to -60°C) and temperature and ice-supersaturation regulate the lifetime of the contrail, which may vary between seconds and hours.[48] Global coverage of line-shaped contrails has been calculated to be 0.1% but up to 5% regionally.[49] Individual studies using satellite data have shown the coverage to be, *e.g.* 0.7% over Western Europe,[50] 0.13% for Thailand,[51] 0.25% for Japan,[51] and 1.3% for the USA.[52] Several estimates of the magnitude of the radiative forcing effect of line-shaped contrails have been made. The estimate of 20 mW m^{-2} (for 1992) presented by the IPCC report[8] is based upon the work of Minnis *et al.* (1999).[53] The forcing is a result of the balance between positive long-wave and negative short-wave effects.[54]

Since the IPCC report,[8] some revised estimates of the forcing have been made, which mostly result in lower forcings, *e.g.* 10 mW m^{-2} and 3.2–10 mW m^{-2}.[55,56] That the more recent radiative forcing estimates are lower is principally because

47 N. Stuber, R. Sausen and M. Ponater, *Theor. Appl. Climatol.*, 2001, **68**, 125.

48 U. Schumann, *Meteorol. Z.*, 1996,**5**, 4.

49 R. Sausen, K. Gierens, M. Ponater and U. Schumann, *Theor. Appl. Climatol.*, 1998, **61**, 127.

50 R. Meyer, H. Mannstein, R. Meerkötter, U. Schumann and P. Wendling, *J. Geophys. Res.*, 2002, D10 10.1029/2001 JD 000426.

51 R. Meyer, R. Büll, C. Leiter, H. Mannstein, S. Marquart, T. Oki and P. Wendling, *DLR-Report No. 176*, Institut für Physik der Atmosphäre, DLR Oberpfaffenhofen, Germay, 2002.

52 P. Minnis, R. Palikinda, J. K. Ayers, D. P. Duda and K. P. Costulis, *Proc. Conf. on Aviation, aerosols, contrails and cirrus clouds, Seeheim, Germany*, ISBSN 92-894-0461-2, European Commission, 2000.

53 P. Minnis, U. Schumann, D. R. Doelling, K. R. Gierens and D. W. Fahey, *Geophys. Res. Lett.*, 1999, **26**, 1853.

54 R. Meerkötter, U. Schumann, D. R. Doelling, P. Minnis, T. Nakajima and Y. Tsuchimia, *Ann. Geophys.*, 1999, **17**, 1080.

55 G. Myhre and F. Stordal, *Geophys. Res. Lett.*, 2001, **28**, 3119.

56 S. Marquart and B. Mayer, *Geophys. Res. Lett.*, 2002, **29**, 10.1029/2001GL014075.

the optical depth of the contrails is lower than previously assumed. An additional GCM study,[57] whilst not reporting radiative forcing values, estimated that contrails might warm the Earth's surface by 0.43°C, a rather large climate response. The recent TRADEOFF estimate (Figure 7) uses 2000 traffic as a base year. For future scenarios, a recent coupled climate simulation has shown that the effects of overall climate change may reduce the effect of contrails in the future because of a warmed upper troposphere, particularly in tropical latitudes.[58] Clearly, there is much more work to do on this subject and the uncertainties remain large.

Cirrus Clouds

The effect of aircraft on cirrus clouds was shown by IPCC[8] (Figure 3) to be potentially large but highly uncertain, such that no best estimate could be given. Aircraft are thought to be able to affect cirrus cloud coverage in two ways: firstly, persistent contrails may spread out to give a cirrus-like cloud coverage (the so-called 'primary effect'); secondly, aircraft emit particles that water in the atmosphere may condense onto (cloud condensation nuclei)—if atmospheric conditions of temperature and humidity change some time after the passage of the aircraft, it is possible that the particles introduced by the aircraft engines may seed cirrus cloud formation (the so-called 'secondary effect'). The primary cirrus effect from spreading persistent contrails is the more demonstrable effect and can lead to extensive cloud coverage that would not have occurred in the absence of air traffic.

Like contrails, cirrus has two opposing effects, of infrared warming and a solar albedo cooling effect, but, overall, thin natural cirrus clouds are considered to warm the Earth's atmosphere whilst thick cirrus clouds cause cooling.[59] There has been a long-term increase of cirrus cloud coverage and although the reasons for this are not understood,[60] it has been suggested that air traffic might be partially responsible.[61]

Two recent independent studies indicate that there is a significant correlation between increases in cirrus cloud coverage and air traffic.[62,63] The estimated increases in cirrus cloud coverage were used within the TRADEOFF project to give an upper limit to the radiative forcing of approximately 100 mW m^{-2}. Independently, Mannstein and Meyer[63] used satellite observations of linear contrails and cirrus clouds, and combined them with an analysis of air traffic density over Europe, which showed that an extra 3% cirrus cloud coverage may be attributable to air traffic. This was estimated to equate to an upper limit for the radiative forcing of approximately 85 mW m^{-2}.

[57] D. Rind, P. Lonergan and K. Shah, *J. Geophys. Res.*, 2000, **105**, 19927.

[58] S. Marquart, M. Ponater, F. Mager and R. Sausen, *J. Climate*, 2003, **16**(17), 2890.

[59] Q. Fu and K. N. Liou, *J. Atmos. Sci.*, 1993, **50**, 2008.

[60] D. P. Wylie and W. P. Menzel, *J. Climate*, 1999, **12**, 170.

[61] O. Boucher, *Nature*, 1999, **397**, 30.

[62] C. S. Zerefos, K. Eleftheratos, D. S. Balis, P. Zanis, G. Tselioudis and C. Meleti, *Atmos. Chem. Phys.*,2003, **3**, 1633.

[63] H. Mannstein and R. Meyer, *Extended Abstracts, Proc. AAC Conference*, Friedrichshafen, Germany, 2003.

Recent estimates of radiative forcing from enhanced cirrus cloud are significant and more work clearly needs to be devoted to this effect to make a more robust estimate.

Sulfate and Soot Particles

As described earlier, aircraft engines introduce particles into the atmosphere from the combustion and in-plume chemical processes. These particles are thought to be of two basic types: volatile (*i.e.* sulfates) and non-volatile (soot). Non-volatile particles of soot are formed inside the combustor in the primary zone and are mostly burned out in further combustion; however, some small particles (of nm aerodynamic diameter size) survive and are emitted from the engine. Sulfate particles are formed from sulfur in the fuel, certainly within the aircraft plume but possibly within the engine itself.

Particles of both forms (volatile and non-volatile) may play a role in the formation of contrails and cirrus clouds but are known to have radiative effects in and of themselves. Sulfate particles backscatter radiation, having a cooling effect; soot particles absorb incident solar radiation and cause local heating. However, both the direct radiative effects of sulfate and soot particles from aviation have been estimated to be rather small. The TRADEOFF estimates[46] have not changed significantly from the IPCC estimates.[8]

5 Reducing the Impacts

Reducing the impact of aviation on climate change represents a major challenge for the coming decades. The industry can be characterized as one that is technologically mature, dominated by long lead-in times from research through to market, long in-service lifetimes, and overwhelming issues of safety to comply with. This, combined with sustained growth rates of the order 3–5% per annum over decadal time frames conspires to make emissions reductions very difficult.

Options for mitigation measures can be broadly placed in three major categories: engine and airframe technological development; operational changes for efficiency and minimizing environmental impact; and future environmentally based approaches. Engine/airframe and ATM fuel efficiencies are overwhelmingly driven by cost-saving potential, which consequentially reduce CO_2 emissions. Future environmentally based approaches may include market-based options for emissions reductions, *e.g.* emissions trading, and more radical operational changes.

The following sections deal with the various approaches outlined above in more detail.

Engine/Airframe Technological Developments

As outlined previously, engine emissions reductions of NO_x, CO, HCs and smoke over the LTO cycle are regulated through ICAO. Currently, of these pollutants, only NO_x seems to remain a real technological and environmental challenge. Although the reduction of these emissions is targeted at the LTO cycle and not for cruise altitudes, technological improvements that reduce NO_x over the LTO cycle generally translate to improvements at cruise. The emissions reductions are negotiated and agreed within ICAO's Committee on Aviation Environmental Protection (CAEP). CAEP 1 regulations took effect in 1986, CAEP 2 has applied to new engines produced since 1996 and CAEP 4 will take effect from 2004. At the time of writing, further stringency is being negotiated within CAEP 6 to be effective from either 2008 or 2012. Reducing NO_x emissions has been far from straightforward as the quest for quieter and more fuel-efficient engines has tended to continually increase overall pressure ratios (OPR). Simplistically, higher temperatures and pressures tend to increase NO_x, such that combustor technology has had to make commensurate improvements to maintain or improve NO_x emissions performance. The regulatory parameter, D_p/F_{oo} [g (NO_x)/kN thrust], varies with engine OPR, such that for higher OPRs a greater D_p/F_{oo} is allowed. Thus, when percentage reductions of NO_x emissions are stated, OPR needs to be considered as the actual emission reduction of a particular engine type may be substantially less, or even increased. Thus, the overall global fleet $EINO_x$ has actually increased from 1992 to 1999 from 12.6 to 13.2, and is forecast to increase to 13.7 by 2015, using a self-consistent set of calculations (References 34, 35 and 37, respectively).

Major research programmes are in place in Europe and the USA to reduce NO_x levels. These programmes focus upon goals well beyond in-service emission reduction capabilities, striving for NO_x reductions in the range of 60 to 70% of today's CAEP 2 regulatory NO_x standard based upon existing LTO certification requirements. The approaches being explored range from combustor sub-component performance improvements to the evolution of new and complex multi-staged combustors. Evolutionary engineering changes to the combustor have minimized increases in NO_x levels of more fuel-efficient engines and, in some cases, have led to reductions of 20% or more of the CAEP 2 standards. More significant reductions, approaching 50% of CAEP 2 standards, are under development and appear attainable within the next decade.[1] Such reductions are being sought through, for example, the optimization of single-stage combustor technology, further improvements in fuel injection, enhanced fuel–air mixing, reduction in combustor liner coolant flow, with more air being made available for combustion, and decreases in hot-gas residence time. Further more detailed overviews of engine technology may be found in the IPCC report[8] and elsewhere.[64]

Improvements in the airframe likewise pose considerable challenges. Laminar flow control, riblets and wing-tip device are all options that require more development. Weight reduction is a more straightforward route to airframe efficiency but this can be offset, in part, as customer demands increase for more complex in-flight entertainment systems *etc.* Radical solutions do exist, such as multi-hulls, tandem wings and canards. The perhaps most familiar revolutionary design is the 'blended wing body'. All these configurations aim to reduce aerodynamic drag and thus operating costs, having the benefit of reducing CO_2 emissions. Nevertheless, many problems need to be overcome and it seems likely that the familiarly styled aircraft will be seen for some decades to come.

A radical longer-term proposal is to develop liquid hydrogen (LH_2) fuelled aircraft. A recent European Commission Research Programme 'CRYOPLANE' has examined the technological feasibility and potential environmental impacts.

If engines were to run on LH_2, they would emit no CO_2 or particles and a smaller amount of NO_x than conventional kerosene engines. The CO_2 produced in the LH_2 production has to be accounted for in an overall environmental assessment. Although virtually no particles would be emitted, theoretical studies show that contrails are likely to form on background particles,[48,65] the aircraft exhaust with the inherent H_2O providing the trigger. A first study of RF of a hypothetical fleet of cryoplanes was made by Marquart *et al.*[65,66] who analysed two scenarios for 2015: the first, a conventional kerosene fleet which stayed constant in size; the second in which the fleet was instantaneously changed from kerosene to LH_2 powered in 2015 and then stayed constant. The instantaneous radiative forcing in 2015 is larger for the LH_2 powered fleet because of the differing physical properties of the contrails formed; however, by 2100, the LH_2 fleet has a smaller radiative forcing than the conventional fleet, because of the long-lasting effects of CO_2 from kerosene combustion. The two scenarios were not intended to represent a realistic situation but facilitate a parametric study. The environmental benefit or otherwise would depend on the scenario adopted and, obviously, the outcome has very large uncertainties. Recently, detailed simulations of contrails from LH_2 powered aircraft have been performed[67] that show that the crystals formed are fewer in number, since they form on the background particles, but are larger than kerosene contrail ice crystals. The consequence of this is that the LH_2 contrails are optically thinner, and potentially have a smaller radiative effect than kerosene contrails. However, because of the larger amount of water vapour emitted with LH_2 powered engines ($2.6 \times$ kerosene fuel), contrails would form over a wider range of temperatures such that total contrail coverage could be greater and, possibly, the total radiative effect larger too—this remains to be investigated.

Emissions Trading

Arguably, the main driver for addressing CO_2 emissions though market-based options (which include emissions trading) is the Kyoto Protocol. The Kyoto Protocol does not currently include international aviation emissions but domestic emissions are accounted for in countries' domestic greenhouse gas inventories (albeit often in a rather crude manner). Nonetheless, the Kyoto Protocol makes allowance for international aviation emissions of CO_2, stating: '*The Parties included in Annex I shall pursue limitation or reduction of emissions of greenhouse gases not controlled by the Montreal Protocol from aviation and marine bunker fuels, working through the International Civil Aviation Organization and the International Maritime Organization, respectively.*' (Article 2, para. 2).

The Kyoto Protocol covers only CO_2 emissions from aviation; NO_x emissions continue to be addressed through ICAO. However, regulation of NO_x emissions (basically as a manufacturing certification standard) is for LTO cycle only. ICAO

[64] J. Lee, in *Towards Sustainable Aviation*, P. Upham, J. Maughan, D. Raper and C. Thomas (eds.), Earthscan Publications, London, 2003.

[65] B. Kärcher, R. Busen, A. Petzold, F. Schröder and U. Schumann, *J. Geophys. Res.*, 1998, **103**, 17129.

[66] S. Marquart, R. Sausen, M. Ponater and V. Grewe, *Aerospace Sci. Technol.*, 2001, **5**, 73.

[67] L. Ström, and K. Gierens, *DLR Report No. 154*, Institut für Physik der Atmosphäre, Oberpfaffenhofen, Germany, 2001.

does not currently address cruise emissions of NO_x in terms of regulation.

For a generalized system of emissions trading to work adequately, it is assumed that permits to pollute represent a market commodity and therefore have an intrinsic value. Also commonly assumed is that the market consists of a broad-based mix of industries and technologies, some technologically mature, others advancing rapidly; some sectors growing, some declining. In such an idealized system, emissions trading is claimed to be a successful approach for reducing emissions. Evidently, careful controls and verification systems must be put in place.

The incorporation of aviation into such a system has been under discussion for the past few years, particularly within ICAO's Committee for Aviation Environmental Protection (CAEP) under the *aegis* of its Working Group 5. Moreover, the EU at the time of writing is proposing a European-wide (general) emissions trading system for CO_2. Several issues require consideration if aviation is to enter into emissions trading. Firstly, aviation is an outlier in the 'idealized' system outlined above, since long-term growth is relatively strong (despite recent events) and it is technologically mature with few opportunities for substantial emissions saving through radical technological innovation. Secondly, aviation has the particular complication exemplified by Figures 4 and 5: the IPCC,[8] showed that aviation's CO_2 emission was 37% of its total radiative effect. The IPCC report,[8] formulated a new metric to illustrate this for comparison with other sectors *etc*., the Radiative Forcing Index (RFI), which is simply the total radiative effect divided by that from CO_2 alone. For 1992, the IPCC[8] showed that this was 2.7 (range 1.9–4.0). The recent TRADEOFF work[46] suggests that the RFI might be greater if cirrus is included. If one assumes that aviation participates in an open emissions trading system, continues to grow and is therefore an overall purchaser of CO_2 permits, then the effective RFI of aviation has the potential to enhance climate change, not reduce it if only CO_2 is considered.[68]

A simple solution to this might be to use the RFI multiplier to increase the cost of permits used for aviation—as an 'exchange rate'. However, there are a number of problems associated with this; most particularly that this could act as a disincentive to further reduce non-CO_2 emissions/effects. Moreover, the RFI is not completely satisfactory in a market situation, since it still encompasses substantial scientific uncertainties. What can be concluded, however, is that the relatively large RFI of aviation cannot be ignored in formulating environmentally credible policy that includes emissions trading to address aviation's climate impacts.

Lastly, if aviation is to enter into emissions trading, a global CO_2 emissions 'cap' is implicit, and thus international emissions need to be attributable to countries. Thus, international aviation emissions need to be allocated in some way so that national CO_2 budgets might be tracked to verify Kyoto commitments. Thus far, an extensive analysis has not been performed. Neither has the ICAO considered this aspect, which underpins the trading regime.

[68] D. S. Lee and R. Sausen, *Atmos. Environ.*, 2000, **34**, 5337.

Operational Measures—'Green Flight'

Operational measures can be considered in two ways; firstly, conventional improvements in increasing the operational efficiency of the air transportation system will minimize fuel burn, decreasing the CO_2 emission per flight. In general, system inefficiencies arise in the air transportation system during taxiing, cruise and landing as a result of air/ground traffic control constraints. Such inefficiencies are the subject of constant improvements. Recently, the vertical separation of air traffic within Europe has been reduced from 2000 feet to 1000 feet. The effect of this is two-fold: the capacity of the system is increased and potentially, aircraft may fly closer to optimally fuel-efficient altitudes. However, by increasing the vertical discretization of traffic, it is conceivable that this will result in *increased* contrail coverage, since there is a greater chance of air traffic flying through ideal contrail conditions.

The second type of operational measure includes more radical solutions such that environmental impacts are minimized by, *e.g.*, avoiding 'environmentally sensitive' parts of the atmosphere. This concept needs some qualification: whilst some of the substantial radiative forcing effects have uncertainties associated with their magnitude, what is more certain is that the mechanisms are reasonably well understood. To explain, the mechanisms for contrail production are well understood and predictable. What is deficient, however, is the adequate quantification of the relevant environmental parameters of the atmosphere, in particular, ice supersaturation. Analysis of in-flight data collected during the 'MOZAIC' experiments has shown that ice supersaturation can display large heterogeneity, ice supersaturated areas being described and characterized as 'moist lenses'.[69] These moist lenses may be quite shallow in depth, less than one flight level (in most areas 2000 ft; in parts of Europe now 1000 ft). Thus, lateral or vertical negotiation of these areas can, in theory, largely avoid contrail production.

Currently, the necessary data are not available from either measurements or predictive models with sufficient accuracy to implement contrail avoidance. Moreover, there are other important issues to be considered: for example, do we know that contrails are worth avoiding? Do we know their radiative effect with sufficient certainty? Can we quantitatively correlate contrails with enhanced cirrus? The answer is clearly in the negative (for a detailed examination of this, see Lee *et al.*[70]), so that implementation of a 'contrail avoidance system' is some considerable time away.

Nonetheless, it is certainly worth exploring the avoidance of this effect with models. The earliest study calculated potential global contrail coverage on a statistical basis, using long-term data from ECMWF.[49] Having calculated potential and actual contrail coverage, air traffic was simply shifted 1 km downwards and upwards to determine the effect. For a downwards shift, contrail coverage increased at northern mid-latitudes because this brings air traffic from the dry lower stratosphere into the moister troposphere but decreased in the tropics as the aircraft were placed in warmer air. More recently, this approach was refined

[69] K. Gierens, U. Schumann, H. G. J. Smit, M. Helten and G. Zangl, *Ann. Geophys.*, 1997, **15**, 1057.
[70] D. S. Lee, P. E. Clare, J. Haywood, B. Kärcher, R. W. Lunnon, I. Pilling, A. Slingo and J. R. Tilston, *DERA report DERA/AS/PTD/CR000103*, DERA Pyestock, 2000.

by simulating air traffic at different real flight level shifts of −2000, −4000 and −6000 feet, recalculating the resultant fuel consumption for the poorer aerodynamics.[71] These calculations were made on-line in a GCM,[72] so that the radiative forcing could be calculated. Again, differential increases and decreases were observed, but at −6000 feet (1.8 km) both contrail coverage and radiative forcing (as annual averages) were substantially reduced, −45% and −47%, respectively, for a 6% increase in fuel. This was found to vary strongly with season and latitude, so that it can be envisaged that a much more refined optimization solution could be found that implies a much smaller fuel penalty.

Shifting cruise altitudes may also affect O_3 radiative forcing. In an initial study, NO_x impacts were considered by Grewe *et al.*[73] who found that decreasing the overall cruise altitude of air traffic by 1 km lowered the O_3 production by 1.5 Tg. Both NO_x and O_3 lifetimes were shorter—radiative forcing changes were not calculated. Using the −6000 feet altitude shift described above, Gauss *et al.*[74] also found a substantial reduction in O_3 production, using a global chemical transport model.

Such 'green flight' studies are in their infancy but it is clear that there are potentially large opportunities for reducing aviation's impact by these more radical changes. In practice, much smaller changes in operation might be required once seasonality and latitude are considered. However, it is stressed that there is much more work to do before implementation into the air transport system could be considered because of the large cost implications to ensure a safe, worthwhile and effective system.

6 Conclusions

Aviation is considered to have a small but significant effect on climate that will increase in both magnitude and share over time.

Aviation affects climate through a number of effects, not just from its CO_2 emissions. These effects include O_3 production in the upper troposphere and lower stratosphere, ambient CH_4 destruction, small direct radiative effects from water vapour and particles, and forcing from linear persistent contrails and enhancement of cirrus clouds. Recent research indicates that the overall radiative forcing may be larger than that estimated by the IPCC in 1999.

Some of the climate effects are still rather uncertain and require more work to make more robust estimates of radiative forcing. However, our ability to predict some of the effects has improved with improved models and data. The most recent research indicates that the largest effects may arise from enhanced cirrus cloud formation although this still has large associated uncertainties.

The possibilities for technologically-driven emissions reductions to ameliorate

[71] C. Fichter, S. Marquart, R. Sausen, D. S. Lee and P. D. Norman, *Extended Abstracts, Proc. of AAC Conference*, Friedrichshafen, Germany, 2003.

[72] M. Ponater, S. Marquart and R. Sausen, *J. Geophys. Res.*, 2002, **107**, 10.1029/2001JD000429.

[73] V. Grewe, M. Dameris, C. Fichter and D. S. Lee, *Meteorol. Z.*, 2002, **11**, 197.

[74] M. Gauss, I. S. A. Isaksen and D. S. Lee, *Extended Abstracts, Proc. of AAC Conference*, Friedrichshafen, Germany, 2003.

climate effects under expected scenarios of growth (of the order 3 to 5% per year) are rather limited. Aircraft engine and airframe technology changes only rather slowly because of long lead-in and in-service times. Currently, the most tangible opportunity is through emissions trading of CO_2 although there are serious conceptual problems with this approach if environmental efficacy is to be assured. Longer-term opportunities exist with altered flight regimes that might significantly reduce some of the impacts. However, much more research is required before such expensive radical options could be considered for implementation.

7 Acknowledgements

This review benefits from participation in projects funded by the European Commission, the Department for Transport and the Department of Trade and Industry, ANCAT/EC2, AERONOX, POLINAT, TRADEOFF, NEPAIR, PARTEMIS, AERO2K. The author also thanks the editors for their encouragement and patience.

Global Warming Consequences of a Future Hydrogen Economy

RICHARD DERWENT

1 Introduction

There is much current interest from scientists, engineers, industry and policy-makers in the role that hydrogen (H_2) may play as a synthetic fuel or energy carrier in the future. Hydrogen is a clean fuel because, when burnt in oxygen, the only combustion product is water. When burnt in air, some oxides of nitrogen may be produced but their formation can be minimized under lean combustion conditions. Hydrogen is seen as having a particularly important future role in the transport sector. Basically, there are two ways to make a vehicle run on hydrogen: by using the hydrogen in an internal combustion engine or by using the hydrogen in a fuel cell. In the latter method, hydrogen is 'burnt' electrochemically in a fuel cell, producing electricity that is then used to drive an electric motor generating the traction. The efficiency of the conversion of chemical energy into mechanical energy is potentially far greater for the fuel cell vehicle compared with the modern internal combustion engined vehicle running on petrol. The combination of an overall efficiency improvement together with a dramatic reduction in pollutant emissions explains the attraction of hydrogen as a future vehicle technology.

Any future hydrogen economy will need to solve the challenging problems involved with the synthesis of hydrogen, its storage and distribution and utilization in all sectors of the energy economy, transport included. Careful environmental assessments will be needed for all of the technologies that are to be employed at each step from synthesis through to utilization to ensure that overall environmental pollution is minimized and that the full potential of hydrogen as a clean fuel is being realised.

Section 2 reviews current understanding of the fate and behaviour of hydrogen in the atmosphere and characterize its major sources and sinks. We show in Section 3 that hydrogen itself, in contrast to most expectations, is a greenhouse gas and we quantify its global warming potential relative to carbon dioxide. In

Issues in Environmental Science and Technology, No. 20
Transport and the Environment

Section 4, we quantify the global warming consequences of replacing the current fossil-fuel based energy economy with one based on hydrogen. The conclusion reached in Section 5 is that unless the leakage of hydrogen from any future hydrogen economy is not carefully controlled there may be little improvement in global warming from the replacement of fossil fuel based energy systems.

2 Fate and Behaviour of Hydrogen in the Atmosphere

Hydrogen (H_2) is a major trace gas in the lower atmosphere or troposphere. The mean global mixing ratio of hydrogen is currently about 510 ppb, 504 ppb in the Northern Hemisphere and 520 ppb in the Southern Hemisphere.[1] Hydrogen is somewhat unusual among trace gases in that although its life cycle has been heavily influenced by human activities, its mixing ratios in the Northern Hemisphere are lower than those in the Southern Hemisphere. This phenomenon is caused by the main sink for hydrogen, surface uptake by soils, which accounts for 80% of the total loss of hydrogen from the atmosphere. Most of the sink, therefore, is over the continental land masses that are concentrated in the Northern Hemisphere.

Recent analyses of long-term observations of hydrogen in the troposphere indicate that mixing ratios have remained fairly constant during the last decade or so. For the Northern Hemisphere, a downwards trend of -2.7 ± 0.2 ppb per year has been reported between 1991 and 1996,[2] a slight upwards trend of $+1.2 \pm 0.8$ ppb per year has been reported based on the continuous observations performed at the Mace Head, Ireland baseline station during 1994–1998, as part of the AGAGE programme.[1] Hydrogen mixing ratios over the entire 1994–2003 period at Mace Head, Ireland[3] show a downwards trend of -0.11 ppb per year.

There have been several previous estimates of the global hydrogen budget. The first analysis,[4] estimated a global production rate of 23.9 Tg yr^{-1}, largely from human activities and a global sink strength of 17.5 Tg yr^{-1}, largely from oxidation by hydroxyl OH radicals:

$$OH + H_2 \rightarrow H + H_2O \tag{1}$$

A global box model has been used[5] to estimate the production of hydrogen from methane CH_4 oxidation, through the photolysis of formaldehyde HCHO:

$$OH + CH_4 \rightarrow CH_3 + H_2O \tag{2}$$

$$CH_3 + O_2 + M \rightarrow CH_3O_2 + M \tag{3}$$

$$CH_3O_2 + NO \rightarrow CH_3O + NO_2 \tag{4}$$

[1] P. G. Simmonds, R. G. Derwent, S. O'Doherty, D. B. Ryall, L. P. Steele, R. L. Langenfelds, P. Salameh, H. J. Wang, C. H. Dimmer and L. E. Hudson, *J. Geophys. Res.*, 2000, **105**, 12,105.

[2] P. C. Novelli, P. M. Lang, K. A. Masarie, D. F. Hurst, R. Myers and J. W. Elkins, *J. Geophys. Res.*, 1999, **104**, 30427.

[3] P. G. Simmonds, S. O'Doherty and G. Spain, *Advanced Global Atmospheric Gases Experiment Mace Head, Ireland. Final Report*, INSCON, 2003.

[4] U. Schmidt, *Tellus*, 1974, **26**, 78.

[5] P. J. Crutzen and J. Fishman, *Geophys. Res. Lett.*, **4**, 321.

Table 1 Globally-integrated sources and sinks for hydrogen

Sources and sinks in Tg yr^{-1}	Seiler and Conrad[6]	*Warneck*[7]	*Novelli et al.*[2]	*Sanderson et al.*[8]
Sources				
Man-made[a]	20	17	15	20
Biomass burning	20	19	16	20
Methane oxidation	15	20	26	15
Oxidation of organic compounds[b]	25	18	14	15
Oceans	4	4	3	4
N_2 fixation	3	3	3	4
Total sources	87	81	77	78
Sinks				
OH-oxidation	8	16	19	17
Soil uptake	90	70	56	58
Total sinks	98	86	75	74

Notes: 1 Tg = 10^{12} g. [a] Hydrogen produced by fuel combustion. [b] See Table 2.

$$CH_3O + O_2 \rightarrow HO_2 + HCHO \quad (5)$$

$$HCHO + \text{radiation} \rightarrow H_2 + CO \quad (6)$$

An additional sink for hydrogen was required to balance these early budgets and this is now known to be the uptake of hydrogen by soils.[4] Subsequently, global hydrogen budgets have been based on the oxidation rates of the primary emitted hydrocarbons and removal by OH-oxidation and uptake by soils.[2,6,7]

A global Lagrangian three-dimensional chemistry transport model (STOCHEM) has been employed to represent the various sources and sinks of hydrogen in the current atmosphere.[8] Global budgets are shown in Table 1. Hydrogen therefore has both natural and man-made sources. Man-made sources include fossil fuel combustion, mainly from petrol-engined motor vehicles through the water gas reaction:

$$CO + H_2O \rightarrow H_2 + CO_2 \quad (7)$$

Biomass burning is another important source, as a by-product of incomplete combustion. Surface ocean waters are generally supersaturated with hydrogen and so act as a small source.[4] Hydrogen is also formed as a by-product of nitrogen fixation in leguminous plants[9] (Table 1).

Significant production of hydrogen also occurs from the oxidation of organic compounds, including methane, by hydroxyl radicals. All these compounds are

[6] W. Seiler and R. Conrad, in *The Geophysiology of Amazonia*, R. E. Dickinson (ed.), John Wiley and Sons, New York, 1987, p. 133.

[7] P. Warneck, in *Chemistry of the Natural Atmosphere*, International Geophysics Series, Academic Press, New York, 1995, Vol. 71.

[8] M. G. Sanderson, W. J. Collins, R. G. Derwent and C. E. Johnson, *J. Atmos. Chem.*, 2003, **46**, 15.

[9] R. Conrad and W. Seiler, *J. Geophys. Res.*, 1980, **85**, 5493.

Table 2 Globally-integrated hydrogen production from the oxidation of organic compounds[8].

Organic compound	*Hydrogen source ($Tg\ yr^{-1}$)*
Methane	15.2
Ethane	0.2
Propane	0.3
Butane	1.2
Ethylene	0.8
Propylene	0.6
Isoprene	11.0
Methanol	0.3
Formaldehyde	0.1
Acetaldehyde	0.1
Acetone	0.5

degraded by OH oxidation through to formaldehyde[10] which undergoes photolysis to produce hydrogen.[11]. Global Lagrangian modelling study[8] shows that the OH-oxidation of methane produces 15.2 Tg yr^{-1} of hydrogen and that of isoprene, 11.0 Tg yr^{-1}, see Table 2. Other organic compounds account for a further 4.1 Tg yr^{-1}.

The main chemical sink for hydrogen in the troposphere in the global Lagrangian modelling study[8] is the reaction with hydroxyl radicals in Reaction (1) above. The largest source of hydroxyl radicals is the reaction of electronically excited oxygen atoms O^1D from the photolysis of ozone O_3, with water vapour H_2O:

$$O_3 + \text{radiation} \rightarrow O^1D + O_2 \tag{8}$$

$$O^1D + H_2O \rightarrow OH + OH \tag{9}$$

The main sink for hydroxyl radicals is the reaction with carbon monoxide CO:

$$OH + CO \rightarrow H + CO_2 \tag{10}$$

The other major sink for hydrogen is uptake by soils at the Earth's surface. Soil uptake rates depend on the nature of the soils, the properties of each trace gas and the rate of transport of the trace gas through the atmospheric boundary layer by turbulence to the Earth's surface. Field studies indicate a dependence of the soil uptake of hydrogen on soil moisture content and on the ecosystem type growing on the soil.[12–14] Measured soil uptake rates, expressed as dry deposition velocities, range from 1.3 mm s^{-1} for savannah systems to 0.1 mm s^{-1} for semi-desert systems. Hydrogen is not removed by uptake on snow, ice, desert or water surfaces. Altogether, soil uptake provides a global sink of 58 Tg yr^{-1}, dominating over OH-oxidation by nearly a factor of four.[8]

For an observed[1] global hydrogen burden of 182 Tg, the global sink strength of 74.4 Tg yr^{-1} from Table I implies an atmospheric lifetime of 2.5 years or

[10] R. Atkinson, *Atmos. Environ.*, 2000, **34**, 2063.
[11] J. G. Calvert and J. N. Pitts, in *Photochemistry*, Wiley, New York, 1967.
[12] R. Conrad and W. Seiler, *J. Geophys. Res.*, 1985, **90**, 5699.
[13] S. Yonemura, S. Kawashima and H. Tsuruta, *Tellus*, 1999, **51B**, 688.
[14] S. Yonemura, S. Kawashima and H. Tsuruta, *J. Geophys. Res.*, 2000, **105**, 14,347.

thereabouts for hydrogen.

3 Hydrogen as a Greenhouse Gas

Any trace gas can be classified as a greenhouse gas if, when present in the atmosphere, it interacts with the incoming solar radiation or with the outgoing terrestrial radiation. Such trace gases are termed direct greenhouse gases or direct radiatively active gases. There are an additional class of trace gases that are not themselves directly radiatively active but they act like radiatively active gases because their presence in the atmosphere perturbs the global distribution of greenhouse gases. These additional classes of trace gases are termed indirect greenhouse gases or indirect radiatively active trace gases.

The emissions of several short-lived tropospheric ozone precursor species exert a profound influence on the urban, regional and global distributions of ozone in the troposphere.[15,16] These tropospheric ozone precursor species include nitrogen oxides (NO_x), methane (CH_4), organic compounds, hydrogen and carbon monoxide (CO). Each of these trace gases has important emission sources from human activities and from natural biospheric processes.[17] Since tropospheric ozone is the third most important greenhouse gas,[18] it follows that an indirect greenhouse effect may be, in principle, associated with the emissions of each of these ozone precursor species because of their potential impact on the tropospheric ozone distribution.[19] Hydrogen, like all ozone precursors, may potentially be an indirect greenhouse gas because its emissions may influence the tropospheric distribution of ozone.

In addition to controlling tropospheric ozone production and destruction, the ozone precursor species also control the tropospheric distribution of hydroxyl radicals[20] and hence the oxidizing capacity of the troposphere. The tropospheric distribution of hydroxyl radicals in turn controls the lifetime and hence global scale build-up of methane,[21] the second most important greenhouse gas after carbon dioxide.[18] There is therefore the potential for the emissions of the ozone precursor gases to alter the tropospheric distribution of hydroxyl radicals and perturb the global scale build-up of methane.[22] Again, hydrogen, like all ozone precursors, may potentially be an indirect greenhouse gas because its emissions may influence the tropospheric distribution of methane.

The importance of hydrogen as a greenhouse gas has been quantified[23] using

[15] P. A. Leighton, in *Photochemistry of Air Pollution*, Academic Press, New York, 1961.

[16] P. J. Crutzen, *Tellus*, 1974, **26**, 47.

[17] J. G. J. Olivier, A. F. Bouwman, C. W. M. van der Maas, J. J. M. Berdowski, C. Veldt, J. P. J. Bloos, A. J. H. Visschedijk, P. Y. J. Zandveld and J. L. Haverlog, *Description of EDGAR version 2.0*, RIVM Report nr. 771060 002, Bilthoven, The Netherlands, 1996.

[18] Intergovernmental Panel on Climate Change, in *Climate Change 2001: The Scientific Basis*. J. T. Houghton *et al.* (eds.), Cambridge University Press, Cambridge, 2001.

[19] R. G. Derwent, *Trace Gases and their Relative Contribution to the Greenhouse Effect*, AERE Report R-13716, H. M. Stationery Office, London, 1990.

[20] H. Levy, *Science*, 1971, **173**, 141.

[21] D. H. Ehhalt, *Tellus*, 1974, **26**, 58.

[22] I. S. A. Isaksen and O. Hov, *Tellus*, 1987, **33B**, 271.

the global Lagrangian chemistry transport model STOCHEM. The model was started from an initial set of trace gas concentrations in October 1994 and used analysed wind fields to run the model through to 1st January 1995. At that point, two model experiments were initiated. The first model experiment continued on without change until 31st December 1998 and this formed the base case. In the second model experiment, the transient case, the hydrogen emission source strength was increased so that a pulse containing an additional 40 Tg of hydrogen was emitted into the model by the 31st January 1995. At this point, the hydrogen emission was reset to the base case value and the model experiment was continued until 31st December 1998. The impacts of the additional hydrogen on the composition of the model troposphere were followed by taking differences between the base and transient cases. These differences in composition between the two experiments were termed 'excess' concentrations. There was no particular significance to the size chosen for the emission pulse and it was given the same spatial distribution as that given to the man-made sources.

A diagrammatic representation of the base and transient case experiments is given in Figure 1, which follows the impact of an emission pulse of hydrogen on the composition of the model troposphere. Figure 1(a) presents the globally-integrated hydrogen emissions over the four years of the experiment, showing the emission pulse of hydrogen during January 1995. The effect of the additional hydrogen emissions is to raise the hydrogen burden in the transient experiment relative to the base case, generating an 'excess' hydrogen burden [Figure 1(b)]. This 'excess' hydrogen decays with an e-folding time of about 2 years, which is close to the atmospheric lifetime of hydrogen.

Because of the increased hydrogen burden in the transient experiment, a decrease is created in the OH burden, which appears as a negative 'excess' in the OH burden, Figure 1(c), through the $OH + H_2$ reaction. Again, the 'excess' OH burden decays with the same e-folding time constant as the 'excess' hydrogen, close to 2 years. The reduced OH burden in the transient experiment, in turn, leads to a decrease in the $OH + CH_4$ reaction flux, Figure 1(d), and the development of a small systematic difference in the globally-integrated methane loss rate from the base case. The time development of the 'excess' $OH + CH_4$ reaction flux follows that shown by the 'excess' OH, decaying with an e-folding time of just under 2 years. Since the $OH + CH_4$ reaction is the major loss process for methane, the hydrogen emission pulse leads to slightly different methane loss rates in the two model experiments and the methane burdens begin to diverge with time. The time development of the 'excess' methane burden resulting from the emission pulse of hydrogen is shown in Figure 1(e). The 'excess' methane burden took about 4 years to reach its maximum before decaying with the time constant associated with the methane adjustment time of about 12 years.

As a result of the increase in the atmospheric burden of hydrogen following the emission pulse, adjustments followed on in the concentrations of all the major tropospheric free radical species and ultimately in tropospheric ozone. Figure 1(f) shows that ozone production was stimulated in the transient case and that an

[23] R. G. Derwent, W. J. Collins, C. E. Johnson and D. S. Stevenson, *Climatic Change*, 2001, **49**, 463.

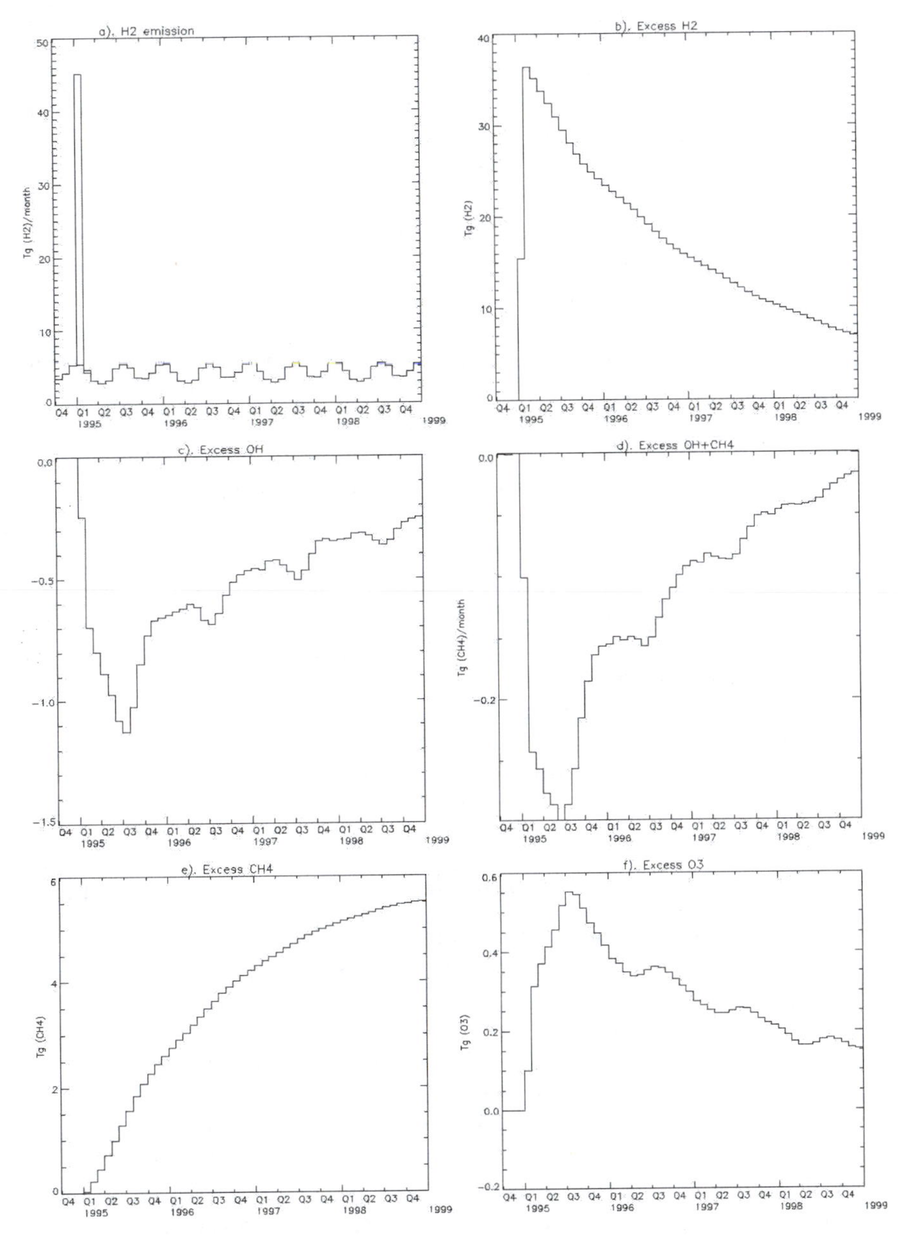

Figure 1 Time development in the composition of the global Lagrangian model troposphere following the emission of a 40 Tg pulse of hydrogen, showing (a) the global hydrogen emissions, (b) 'excess' H_2 burden, (c) 'excess' OH burden, (d) 'excess' OH + CH_4 reaction flux, (e) 'excess' CH_4 burden, and (f) 'excess' O_3 burden

'excess' ozone burden quickly developed following the emission pulse. The 'excess' ozone burden decayed with an e-folding time of about 2.5 years, close to the atmospheric lifetime of hydrogen.

Using the literature radiative forcing formulae,[24] the time-integrated 'excess'

[24] Intergovernmental Panel on Climate Change, *Climate Change: The IPCC Scientific Assessment*, J. T. Houghton *et al.* (eds.), Cambridge University Press, Cambridge, 1990.

methane burdens in Figure 1(e) were converted into time-integrated radiative forcing, over a 100-year time horizon. The 1 Tg hydrogen emission pulse produced a time-integrated methane radiative forcing of 0.35 mW m^{-2} year over a 100-year time horizon. The radiative forcing consequences of the tropospheric ozone burden changes were evaluated using an appropriate radiation code.[25] The 1 Tg emission pulse produced a time-integrated ozone radiative forcing of 0.25 mW m^{-2} year over a 100-year time horizon.

These time-integrated radiative forcings were converted into Global Warming Potentials (GWPs) by comparison with the time-integrated radiative forcing of a reference gas, usually taken to be carbon dioxide, CO_2. Here we define the Global Warming Potential of hydrogen as the ratio of the time-integrated radiative forcing for a particular radiative forcing mechanism (whether through methane or ozone changes) resulting from the emission of 1 Tg of hydrogen compared with that from the emission of 1 Tg of CO_2 over a 100-year time horizon. The fate of a 1 Tg emission pulse of CO_2 was described using the CO_2 response function of the Bern carbon cycle model.[26] On this basis, the GWP^{CH4} and GWP^{O3} for hydrogen were found to be 3.4 and 2.4, respectively, over a 100-year time horizon. The overall GWP for hydrogen is therefore 5.8 over a 100-year time horizon.

It is a clear consequence of the global chemistry transport modelling[23] that, due to the emission of hydrogen to the troposphere, changes occur in the global distributions of methane and ozone, the second and third most important greenhouse gases. Emissions of hydrogen lead to increased burdens of methane and ozone and hence to an increase in radiative forcing. Hydrogen is clearly an indirect radiatively active greenhouse gas with a global warming potential of 5.8 over a 100-year time horizon.

4 Greenhouse Gas Consequences of a Global Hydrogen Economy

The emission of hydrogen to the troposphere leads to changes in the global distributions of methane and ozone, leading to increased radiative forcing. This prompts the questions as to whether a global hydrogen economy would have consequences for global warming and, ultimately whether a hydrogen economy would be better or worse than the fossil-fuel economy it replaces.

The global hydrogen production capacity required to replace the entire current fossil-fuel based energy system is estimated to be of the order of about 2500 Tg H_2 yr^{-1}. If there was a leakage rate of the order of say 1%, then the global hydrogen economy would emit about 25 Tg H_2 yr^{-1}. Using the GWP of 5.8 described in Section 3, the global hydrogen economy would have the radiative forcing equivalent to 25×5.8 Tg CO_2 yr^{-1}, that is to say about 150 Tg CO_2 yr^{-1}. The fossil fuel system it would replace has a CO_2 emission of 23 000 Tg CO_2 yr^{-1}. On this basis, the global hydrogen economy with a leakage rate of 1% has a climate impact of 0.6% of the fossil fuel system it replaces. If the leakage

[25] J. M. Edwards and A. Slingo, *Q. J. R Meteorol. Soc.*, 1996, **122**, 689.

[26] Intergovernmental Panel on Climate Change, *Climate Change 1995: The IPCC Scientific Assessment*, J. T. Houghton *et al.* (eds.), Cambridge University Press, Cambridge, 1996.

rate was 10%, then the climate impact would be 6% of that of the fossil fuel system.

5 Conclusions

Hydrogen-based energy systems appear to be an attractive proposition in providing a future replacement for the current fossil-fuel based energy systems. Hydrogen appears attractive because it is a clean fuel and because it offers efficiency improvements when it is utilized. The transport sector may provide some of the first applications of the novel hydrogen technologies.

Hydrogen is an important, though little studied, trace component of the atmosphere. It is present at a mixing ratio of about 510 ppb currently and has important man-made and natural sources. It atmospheric lifetime is about 2.5 years and there is a global burden of about 180 Tg in the atmosphere.

Because hydrogen reacts with tropospheric hydroxyl radicals, emissions of hydrogen to the atmosphere perturb the distributions of methane and ozone, the second and third most important greenhouse gases after carbon dioxide. Hydrogen is therefore an indirect greenhouse gas with a global warming potential GWP of 5.8 over a 100-year time horizon. A future hydrogen economy would therefore have greenhouse consequences and would not the free from climate perturbations.

If a global hydrogen economy replaced the current fossil fuel-based energy system and exhibited a leakage rate of 1% then it would produce a climate impact of 0.6% of the current fossil fuel based system. If the leakage rate were 10%, then the climate impact would be 6% of the current system.

Careful attention must be given to reduce to a minimum the leakage of hydrogen from the synthesis, storage and utilization of hydrogen in a future global hydrogen economy if the full climate benefits are to be realised in comparison with fossil fuel based energy systems.

6 Acknowledgements

The author is grateful to the Department for Environment, Food and Rural Affairs for their support under contract no. EPG 1/3/164, to Professor Peter Simmonds of the University of Bristol for providing access to the observational record for hydrogen at Mace Head, Ireland, to Michael Sanderson of the Met Office for his help with the modelling of the life cycle of hydrogen, to Bill Collins and Colin Johnson for their help with the modelling of the fate and behaviour of the hydrogen emission pulses and for David Stevenson of the University of Edinburgh for his help with the calculation of radiative forcings.

Sustainable Transport and Performance Indicators

HENRIK GUDMUNDSSON

1 Introduction

Since the release of the report of the World Commission on Environment and Development (known as the Brundtland Report) in 1987 transport problems and policies have increasingly been framed with regard to the notion of Sustainable Development. According to the Brundtland Commission and many subsequent statements sustainable development refers to '. . . *development that meets the needs of the present without compromising the ability of future generations to meet their own needs*'.[1] The relevance of this notion for transport has chiefly been inferred from the fact the movement of people and goods serves present society while contributing to a range of pressures on the environment, from the impairment of air quality at urban and street level to the emissions of greenhouse gasses at the global scale.[2] Moreover, current and future expected growth patterns of transport appear to be at odds with what can in the long run be sustained by limited environmental and economic resources and capacity.[3] While these concerns are not entirely new, the political prominence given to 'sustainability' at the international level appears to have forced governments and others to address the full range of transport impacts in a more integrated way than before, and also to reconsider the very role of transportation in the pursuit of further economic and social development. Despite variations in emphasis a new policy agenda calling

[1] World Commission on Environment and Development, *Our Common Future*, 1987, Oxford University Press, Oxford.

[2] D. L. Greene, Sustainable Transportation, in *The International Encyclopaedia of the Social and Behavioral Sciences*. P. B. Baltes and N. J. Smelser (eds.) Elsevier Science, Oxford, 2001, pp. 15335–15339.

[3] US Transportation Research Board. *Toward a sustainable future. Addressing the Long Term Effects of Motor Vehicle Transportation on Climate and Ecology*. Transportation Research Board, 1997, Special report 251, National Academy Press, Washington D.C.

Issues in Environmental Science and Technology, No. 20
Transport and the Environment

for more 'Sustainable Transport' or 'Sustainable Mobility' has shaped national and local policy development in several countries, not least in the European Union.[4]

However, while it is most likely that transport will remain important for both the environment and society it has not been altogether clear what 'Sustainable Transport' would imply. Many questions have been raised, both from a theoretical point of view and as more practical concerns:[5] First, what should be sustained, more precisely? Is it the transport systems, as we know them today or is it rather the services in terms of access and opportunity they provide? Sustainable in what respect one may ask next? It makes a difference if the perspective is a decade or a century, and if the context is a city or the entire globe. Furthermore, environmental protection may not be the only relevant concern; economic functions of transport may also have to be considered. Even in addressing the environmental dimension there are fundamental questions: Which are the critical ecological or health based limits to be observed and what is the role of transport in transgressing them? In other words it may not be entirely clear what kind of requirement a 'sustainable transport system' should fulfil, how far away the present systems are from satisfying them, and how policies can help to govern development in the desired direction.

Questions such as these have provoked a need for operational tools to navigate in an increasingly complex world. Among the most popular tools are *indicators* and *performance measures*. Indicators are selected variables that can help to make objectives operational and reduce the complexity in dealing with system management and intervention. They can function as guideposts in technical analysis and policy making as well as for the general public debate. When indicators are compared with standards or objectives they become performance measures, measuring the performances of systems, organizations or policies. Indicators and measures of transport have been incorporated into many different kinds of monitoring and assessment frameworks. Examples include transport sections in general sustainable development indicator frameworks such as the UK's *Quality of Life Counts*,[6] environmental indicator sets such as the 'Environmental Signals' from the European Environment Agency[7] and many similar initiatives in a local context.[8] However, more specific frameworks have also been set up to monitor or forecast the performance of transport systems or policies at

[4] ECMT, *Sustainable Transport Policies*. European Conference of Ministers of Transport, Paris, 2000 and CEC, *The Future Development of the Common Transport Policy—a Global Approach to the Construction of a Community Framework for Sustainable Mobility*, Commission of the European Communities, Brussels, 1993.

[5] See *e.g.* OECD, Towards sustainable transportation, *The Vancouver Conference, OECD Proceedings*, Organisation for Economic Co-operation and Development, Paris, 1997, and H. Gudmundsson H and M. Höjer. Sustainable development principles and their implications for transport, *Ecolog. Econom*, 1996, **19** 269-282.

[6] DETR, *Quality of Life Counts. Indicators for a Strategy for Sustainable Development for the United Kingdom: A Baseline Assessment*, Department of the Environment, Transport and the Regions, London, December 1999.

[7] EEA, *Environmental signals* 2001, European Environment Agency regular indicator report, Environmental assessment report No. 8, European Environment Agency, Copenhagen, 2001.

[8] See *e.g.* Y. Rydin, *Indicators Into Action: Local Sustainability Indicator Sets in Their Context*, The Pastille Consortium, London School of Economics, London, 2002.

both European, national and local levels.[9] Within such frameworks dedicated indicators to monitor 'sustainability' aspects of transport have sometimes been incorporated, and researchers and experts have even devised whole systems to specifically monitor sustainable development in a transport context.[10]

Broadly speaking the literature represents three different approaches to make Sustainable Transport operational and measurable using indicators:

- In the first approach 'Sustainable Transport' serves as a metaphor of a broad policy agenda where transport policies take into account (also) sustainable development concerns. Policy planning and evaluation in this context typically incorporate some relevant indicators, such as growing transport volumes or carbon dioxide emissions from transport.[11]
- In the second approach 'Sustainable Transport' is taken literally as meaning transport that can be sustained given certain limitations in time and space set by the environment and/or by certain demands of society. This approach derives from explicit reflections over the meaning of sustainability and what measuring it would entail in the more limited context of transport.
- The third approach represents a mixture of the above, in which 'literal' explorations of the sustainability concept are used to guide the construction of indicators that can inform either research or policy assessment subscribing to the 'Sustainable Transport' agenda.

While the first, metaphorical approach represents a typical stance adopted by many policy administrations throughout the world, the second, literal, one has mostly been pursued by some academics. In the third approach this is taken further by researchers or experts to more directly support, assess or critically examine transport trends or policies.

The present review focuses mostly on the second and third approach. This should not suggest that the numerous contributions in the policy realm are irrelevant. Rather, the scope in this chapter is limited to the more substantial conceptual contributions, thus aiming to provide an overview of how sustainability of transport may be measured and monitored using various kinds of performance indicators, as well as discussing some of the experience and implications of using them to support or assess policy making. Section 2 will set out overall aspects of sustainable transport indicators treating in turn conceptualization, operationalization and utilization issues. Section 3 will review a limited number of existing or proposed indicator sets of various origin. Section 4 concludes with a consideration of what may be inferred concerning the sustainability of transport systems from sustainable transport performance indicators and measurement frameworks.

[9] H. Van der Loop, *Transport Policy Monitoring in Europe. Proceedings of the Workshop held in Amsterdam, 17–18 October 2002*, Ministry of Transport, Public Works and Water Management, AVV Transport Research Centre, Rotterdam, April 2003.

[10] See examples later in this chapter or, for instance, R. Gilbert, N. Irwin, B. Hollingworth and P. Blais, *Sustainable Transportation Performance Indicators (STPI)*, Project Report on Phase 3, The Centre for Sustainable Transportation, Toronto, 31 December, 2002.

[11] Prominent representative of this approach are: 1993 European Commission White Paper *The Future Development of the Common Transport Policy—a Global approach to the construction of a Community framework for sustainable Mobility* and: UK Department for Transport, *A New Deal for Transport. Better for Everyone*. The Stationery Office, London, 1998.

2 Making Sustainable Transport Operational

The use of indicators to measure and monitor sustainable transport involves several tasks. Drawing from the indicator literature[12] we will address three components of this process. The first component is *conceptualization*, which defines what is to be monitored, in this case sustainable development aspects of transport. The second component is *operationalization* in which concepts are made measurable by selecting parameters and indicator types. The third component is *utilization*, which refers to the ways in which the indicators are drawn upon in analysis or policy. In principle there are strong relations between these steps. The intended use should, for instance, influence the concepts to be specified, while indicator selection will again restrict possible uses (as when the choice of only quantitative emission data as environmental indicators may disregard concern for the more complex dimensions of urban liveability). In practice, however, there is not always a clear line from concept to measurement to use.

Conceptualization

As already indicated, sustainable transport is a somewhat nebulous concept, and various sources of inspiration have generally been drawn upon to specify it. We can distinguish between normative, analytical and strategic inspirations.[13] One key inspiration is of course the international debate on sustainable development in general and the various normative concepts offered on that scene. This includes the 'Brundtland' concern for the well being of both present and future generations, as well as the specific attention given to maintenance of Earth's life-support systems. Another influential notion drawn from this debate is the three-dimensional, or 'triad',[14] approach claiming that sustainable development must encompass economic, social and environmental dimensions (Figure 1). A fourth dimension is sometimes added, referring to institutions governing trade-offs or synergies within the three others.[15]

The above notions are present in practically all documented attempts to conceptualize sustainable transport. However, the emphasis put on various dimensions differ. While some explicitly address only 'sustainability' (effectively concerns for future generations, such as maintaining resources), most contributions include also 'development' aspects (transport outcomes of interest to the present generation, such as mobility, noise, accidents, *etc.*). Similarly, there is a division between those contributions that bridge all three dimensions *versus* the ones emphasizing the environmental dimension, often specified as 'Environmentally

[12] J. I. De Neufville, *Social Indicators and Public Policy: Interactive Processes of Design and Application*, Elsevier, Amsterdam, 1975.

[13] E. Becker, T. Jahn, I. Stiess, and P. Wehling. *Sustainability: A Cross-Disciplinary Concept for Social Transformations*, 1997. Most Policy Papers no 6, UNESCO, Paris.

[14] Expression from: N. Low, Is Urban Transport Sustainable? in *Making Urban Transport Sustainable*, ed. N. Low and B. J. Gleeson, 2003, pp. 1–21. Palgrave, Basingstoke.

[15] Agenda 21, the key policy document endorsed at the United Nations Conference on Environment and Development in Rio de Janeiro in 1992, is often considered a key reference for the four dimension view. In the sustainable transport literature the three first dimensions (economic, social, environmental) are most often referred to.

Figure 1 Dimensions of sustainable development

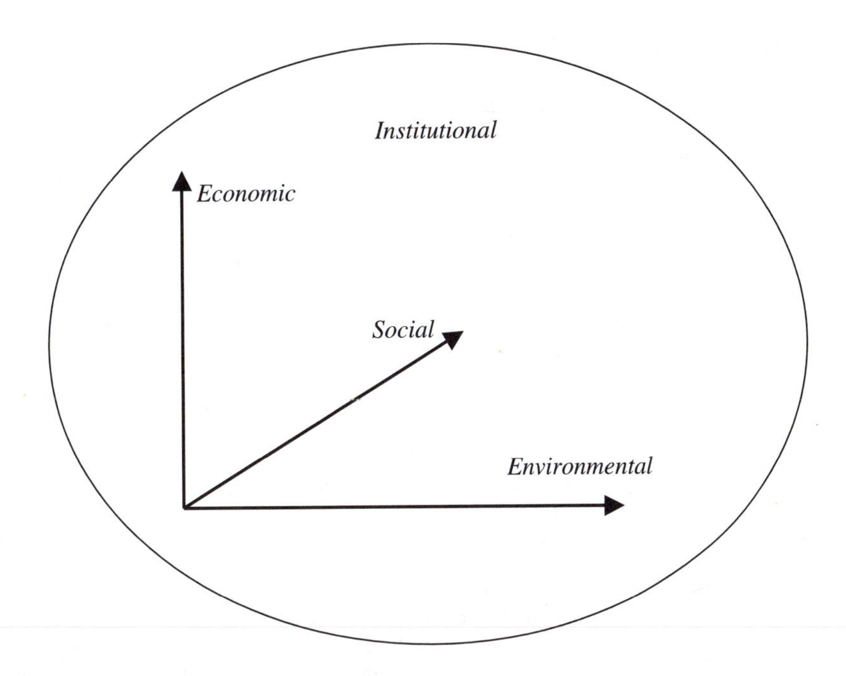

Sustainable Transport'.[16] Table 1 suggests a grouping into these dimensions of selected key references from the conceptual literature on sustainable transport.

Secondly, different academic disciplines offer important analytic building blocks to set up more specific sustainability concepts, criteria and metrics. Environmental sciences have provided influential notions such as Carrying Capacity, and Critical Loads and have also formed the basis for more detailed assessment tools such as emission inventories, air quality models and Global Warming Potential index, all of which have been widely incorporated in sustainable transport indicators. Environmental economics has contributed several important ideas. Most fundamentally perhaps by defining sustainability in terms of preservation of society's capital base, often divided into so-called 'weak' and 'strong' notions of sustainability.[17] Widely cited in the sustainable transport literature are the 'strong sustainability' rules proposed by ecological economist Herman Daly, according to which: (1) renewable resources should not be used faster than their regeneration rates, (2) non-renewable resources should not be used faster than substitutes become available, and (3) pollution should not

[16] As for instance the OECD project with this title (OECD-PPCG *Environmental Criteria for Sustainable Transport. Working Paper of the PPCG Task Force on Transport*, Pollution Prevention and Control Group, Organisation for Economic Co-operation and Development, Paris, 1996).

[17] The 'weak' notion suggests that economic assets can be substituted for environmental ones, further implying that environmental assets should be measured in monetary terms, while 'strong' sustainability suggest non-substitution leading to a need for separate accounting principles defined by biophysical conditions and limits, see *e.g.* R. K. Turner (ed.) *1993 Sustainable Environmental Economics and Management Principles and Practise.* Belhaven, London. In practice a 'strong' view is most often implied if a range of indicators of sustainable transport are proposed, as opposed to aggregating results in one figure such as discounted costs.

Table 1 Approach to sustainable transport in selected references

	Environmental	*3 dimensions*
Sustainability (future generations)	a	
Sustainable development (present and future generations)	b–d	e–p

[a] US Transportation Research Board, *Toward a Sustainable Future. Addressing the Long Term Effects of Motor Vehicle Transportation on Climate and Ecology*, Transportation Research Board, Special report 251, National Academy Press, Washington D.C, 1997.
[b] Royal Commission on Environmental Pollution, *Transport and Environment. Eighteenth report. Cm* 2674, HMSO, London, 1994.
[c] P. Kageson, *The Concept of Sustainable Transportation*, The European Federation for Transport and Environment, T&E 94/3, Bruxelles, 1994.
[d] OECD-PPCG, *Environmental Criteria for Sustainable Transport. Working Paper of the PPCG Task Force on Transport*, Organisation for Economic Co-operation and Development, Paris, 1996.
[e] J. Whitelegg, *Transport for a sustainable future. The case for Europe*, Belhaven Press, London, 1993.
[f] UK Round Table on Sustainable Development, *Defining a Sustainable Transport Sector*, UK Round Table on Sustainable Development, London, 1996.
[g] H. Gudmundsson, & M. Höjer, Sustainable development principles and their implications for transport, *Ecol. Econom.*, 1996, **19**, pp. 269-282.
[h] W. Spillmann, *et. al.* Criteria for sustainable mobility. (English summary from): *Nachhaltigkeit: Kriterien im Verkehr*, Ernst Basler & Partner, Zollikon, Switzerland, 1998.
[i] Wuppertal Institute *et al.*, *Baltic* 21 *Transport Sector Report*. Baltic 21 Series No 8/98, UBA-TEXTE 38/98, Federal Environmental Agency (Umweltbundesamt), Berlin, 1998.
[j] Joint Expert Group on Transport and Environment, *Recommendations for Actions for Sustainable Transport, A strategy Review* Commission of the European Community Directorate-General Transport & Directorate-General Environment, Brussels, 26, September 2000.
[k] W. R., Black. Toward a measure of transport sustainability, *Transportation Research Board Meeting, 2000, Conference Preprints*, Transportation Research Board, Washington, D.C., 2000.
[l] E. Akinyemi and M. Zuidgeest, *Sustainable Development & Transportation: Past Experiences and Future Challenges*, World Transport Policy & Practice, 2000, Vol. 6, No. 1, pp. 31–39.
[m] D. L. Greene, Sustainable Transportation, in P. B. Baltes & N. J. Smelser (eds.), *The International Encyclopedia of the Social & Behavioral Sciences.* Elsevier Science, Oxford, 2001, pp. 15335–15339.
[n] H. Minken, A framework for the evaluation of urban transport and land use strategies with respect to sustainability. Paper presented to the *Sixth Workshop of the Transport, Land Use and Environment (TLE) Network*, Haugesund, September 27–29, 2002.
[o] T. Litman and D. Burwell, *Issues in Sustainable Transportation*, Victoria Transport Policy Institute, Victoria, British Columbia, 2003.
[p] N. Low, Is Urban Transport Sustainable? in, N. Low and B. J. Gleeson (eds.), *Making Urban Transport Sustainable*, Palgrave, Basingstoke, 2003, pp. 1–21.

exceed the assimilative capacity of the environment.[18] Assessments of transport directly based on such rules often suggest that current transport trends are unsustainable.[19] Other highly influential economic ideas in the sustainable transport literature include *eco-efficiency* (referring to an increase in economic output without an equivalent increase in environmental damage)[20] and the need to correct for *market failures* (*e.g.* the need to internalise external environmental costs in transport prices).[21]

Third, and finally, sustainable transport notions have been highly influenced by strategic issues, both in terms of policy recommendations from the international sustainable development process, and in terms of more specific issues on current transport policy agendas. The former has injected notions such as the call for closer integration of economic and ecological reasoning in decision-making,[22] and the need to actively engage citizens in policy processes. The latter involve critical transport issues such as shifting to new technologies and alternative energy carriers, optimizing the logistical organization of transport flows, and mitigating urban congestion, air pollution and noise. All of these strategic topics have been flagged in various connections as key levers to more sustainable transport. In fact some scholars even claim that the only way to make sustainable transport operational is to explore the feasibility of practical policy options, rather than engaging in theoretical reconstruction.[23]

What should emerge from the above is that there are many sources of inspiration behind the concept of sustainable transport, all of which do not readily assemble into a uniform idea. No common agreement therefore exists on a specific meaning of the term.[24] Among the key factors contributing to obscure the notions are:

- The idea of sustainable development itself is contested, normative and multi-dimensional, a wide range of methodologies and metrics being applied to measure various aspects of it.
- The transport sector, consisting of many technical and social subsystems

[18] H. Daly, Toward some operational principles of sustainable development, *Ecolog. Econom.*, 1990, **2**, 1–6.

[19] See for instance: J. S., Szyliowicz, Decision-making, intermodal transportation, and sustainable mobility: towards a new paradigm. *Int. Social Sci. J.*, June 2003, **55**, (2), 185–197, and H. Gudmundsson and M. Højer, Sustainable development principles and their implications for transport. *Ecolog. Econom.* 1996, **19**, 269–282.

[20] R. Montgomery, and L. Sanches, Efficiency: The sustainability criterion that provides useful guidance for statistical research, *Statistical J. UN ECE*, 2002, **19**, 29–40.

[21] As for instance in the following proposition: 'Having a sustainable transport system means making each road user pay at least the full marginal cost of his or her journey'. (D. Maddison, O. Johansson and D. Pearce. *The True Costs of Road Transport*, Blueprint, nr. 5. Earthscan, London, 1996, p. 146.)

[22] Environmental integration is a highly influential notion in the policies of the European Union, as for instance reflected in Article 6 in the EC Treaty: 'Environmental protection requirements must be integrated into the definition and implementation of the Community policies and activities (...) in particular with a view to promoting sustainable development'. (The Treaty on European Union and the Treaty establishing the European Community, *Official J. Eur. Communities*, C 325/1 24.12.2002).

[23] E. T. Verhoef and E. Feitelson, (eds.) *Transport and Environment: in Search of Sustainable Solutions*, Edward Elgar, Cheltenham, 2001.

[24] Greene 2001 (Reference 2).

interacting to produce social benefits as well as negative environmental effects.
- Transport not being isolated from the rest of society, meaning that sustainability of transport systems should in fact be considered as part of changes in the whole socio-economic system.

Nevertheless the predicament of sustainability can hardly be removed from the context of transport simply because it involves a number of complications. Moreover, the notion of sustainable transport has been conceptualised quite extensively in various forms. The most straightforward form is by *defining* it. Other forms include the statement of *criteria*, *principles*, *objectives* or *targets*. Most attempts can be summarized into a number of *issues* considered to be seminal to the concept. We will look into a few examples.

Definitions. Definitions of sustainable transport or mobility proposed in the literature chiefly extend the Brundtland concept mentioned above. One definition simply suggests that '*... sustainable transport is satisfying current transport needs without jeopardising the ability of future generations to meet these needs.*'[25] In the United Kingdom the former Round Table of Sustainable Development similarly proposed that a sustainable transport policy '*... seek[s] to minimise current and anticipated future adverse impacts and their associated costs, while continuing to deliver or improve existing benefits.*'[26] Definitions in terms of what kind mobility would be worth sustaining (rather than what factors would limit it) are more rare.[27]

Other definitions are focussed on the environmental aspects. An OECD project defined in 1996, Environmentally Sustainable Transport as '*... transportation [that] does not endanger public health or ecosystems and meets needs for access consistent with (a) use of renewable resources below their rates of regeneration, and (b) use of non-renewable resources below the rates of development of renewable substitutes.*'[28] This influential contribution (drawing from Daly above) has been combined with other dimensions of sustainability by the Canadian Centre for Sustainable Transportation[29] and The European Commissions Expert Group on Transport and Environment[30] leading to a comprehensive definition of a sustainable transport system as one that:

- 'Allows the basic access and development needs of individuals, companies and

[25] W. R. Black, Socio-economic barriers to sustainable transport. *Transport Geog.*, 2000, **8**. (p. 141). This definition implies that sustainability of current transport trends are only considered in terms of their effects on future *transport* needs, a somewhat narrow perspective. As we shall see in Section 3 the same author also offers a broader approach.

[26] UK Round Table on Sustainable Development: *Defining a sustainable Transport Sector.* UK Round Table on Sustainable Development, London, 1996.

[27] One example would be the concept of 'customised mobility' (R. Kemp & J. Rotmans, *Transition Management for Sustainable Mobility.* MERIT, the Maastricht Economic Research Institute on Innovation and Technology of Maastricht University, Maastricht, 17 January, 2002)

[28] OECD, *Policy Instruments for Achieving Environmentally Sustainable Transport*, Organisation for Economic Co-operation and Development, Paris, 2002. One notes the reflection of Daly's propositions, see Reference 18.

[29] see URL: www.cstctd.org/index.html

[30] Joint Expert Group on Transport and Environment, *Recommendations for Actions for Sustainable Transport. A Strategy Review.* Commission of the European Community Directorate-General Transport & Directorate-General Environment, Brussels, 26 September 2000.

societies to be met safely and in a manner consistent with human and ecosystem health, and promotes equity within and between successive generations;
- Is affordable, operates fairly and efficiently, offers choice of transport mode, and supports a competitive economy, as well as balanced regional development;
- Limits emissions and waste within the planet's ability to absorb them, uses renewable resources at or below their rates of generation, and, uses non-renewable resources at or below the rates of development of renewable substitutes while minimising the impact on the use of land and the generation of noise.'[31]

While these definitions may reflect overall dimensions of sustainable development as well as a range of transport policy concerns they do not provide criteria for more rigorous assessment of the sustainability of any particular transport situation or decision. Definitions of a more technical kind have been proposed but tend to have much more limited applications.[32]

Principles of sustainable transport refer to criteria to be followed in dealing with transport systems or policies. In the 'system principles' perspective transport is considered in regard to the threats they pose to overall sustainability principles such as resource conservation rules or principles concerning positive contributions such as economic efficiency. On the basis of such principles objectives and strategies may be proposed. In the second, 'policy principles' perspective, the starting point is taken in strategic governance principles that need to be adopted to influence transport development, such as internalisation, participation or integration. If these principles are adhered to, a sustainable transport situation should ensue. A prominent example is the so-called Vancouver principles of sustainable transportation (combining both of the above) adopted at an OECD conference in 1996.[33]

Targets of sustainable transport are quantitative measures of the reduction in transport volume or its impacts required for transport systems to fulfil sustainable development definitions, principles or criteria. While such targets may provide rigorous measures for assessment, it is nevertheless difficult to establish them based on empirical reference. Besides the problems involved in the definition of absolute environmental thresholds at a system level in general, there is the additional problem of allocating shares or reduction burdens across sectors. Different principles for such an allocation may be proposed,[34] but few attempts have been made to apply them fully in practice. The targets that have been

[31] This formulation has even been officially adopted by the Transport Ministers of the European Union: Council (Transport/Telecommunications) Strategy for integrating environment and sustainable development into the transport policy, *Council Resolution 2340*, Council Meeting, Luxembourg (4/4/2001) Press:131 Nr: 7587/01.

[32] For a definition in an urban planning context see, *e.g.*, H. A. Minken, Framework for the evaluation of urban transport and land use strategies with respect to sustainability. Paper presented to the *Sixth Workshop of the Transport, Land Use and Environment (TLE) Network*, Haugesund, 27–29 September, 2002. In a road traffic-modelling context see A, Nagurney, *Sustainable Transportation Networks*, Edward Elgar, Cheltenham, 2000.

[33] OECD, *Towards Sustainable Transportation. The Vancouver Conference, OECD Proceedings*, Organisation for Economic Co-operation and Development, 1997, Paris.

[34] See, *e.g.* P. Nijkamp and J. Vleugel, *in search of sustainable transport systems*, in D. Banister, R. Capello, P. Nijkamp, (eds.), *European Transport and Communications Networks. Policy Evolutions and Change*. John Wiley & Sons, Chichester, 1995, pp. 278–299.

Table 2 Sustainable transport targets proposed for Europe (D. Banister, D. Stead, P. Steen, J. Äkerman, K. Dreborg, Nijkamp and R. Schleisher-Tappeser, *European Transport Policy and Sustainable Mobility*, Spon Press, London and New York, 2000)

Environmental targets
25% reduction of CO_2 emissions from 1995 to 2020
80% reduction of NO_x emissions from 1995 to 2020
No degradation of specially protected areas
Minor (2%) increase of net infrastructure surface in Europe
Regional development targets
Improve relative accessibility of peripheral regions (both internal and external)
Efficiency targets
Full cost coverage (including external costs) of transport under market or equivalent conditions
Reduce public subsidies to all forms of transport to zero

suggested in a sustainable transport context mostly represent either political compromise or pragmatic notions for exploratory research. An example of the latter is shown in Table 2. In both counts targets may be very valuable tools (as discussed further in relation to performance indicators below), but the degree to which they reflect actual limits, dividing possible transport situations into sustainable and unsustainable ones, may often be questioned.

Issues. The most widely applied approach to preparing the concept for operationalization is simply to list the major tangible problems or issues it is assumed to encompass. This typically includes environmental problems such as acidification or global warming, economic issues such as transport infrastructure investments or social issues such as access for all. The route towards establishing the list of issues may proceed from bringing overall sustainability definitions, principles or issues down to bear on the transport sector ('top-down'), or conversely by confronting a review of current transport problems with possible implications in terms of sustainable development ('bottom-up'). Either way there is scope for pragmatism since there is no generally accepted procedure for the application of general ideas of sustainable development to individual sectors. Table 3 lists the major issues of sustainable transport raised in several studies.

System Boundaries. An important complication in the conceptualization process is drawing a system boundary. In the examples above a general notion of the transport sector including transport by all modes is often assumed. This is at odds with standard national accounting procedures that exclude own-account and private transport from the sector per se. Another possibility is to also include production and disposal of transport system components. In any case the boundary drawn between transport and non-transport systems may be challenged. Often, focus is on a sub-sector or sub-system level where notions such as 'Sustainable Urban Travel',[35] 'Sustainable Road Transport',[36] 'Sustainable Supply

[35] ECMT, *Implementing Sustainable Urban Travel Policies. Final report*, European Conference of Ministers of Transport, ECMT, Paris, 2002.

[36] J. Schwaab and S. Thielmann, *Economic Instruments for Sustainable Road Transport—An Overview for Policy Makers in Developing Countries*, Deutsche Gesellschaft für Technische

Table 3 Overview of main issues of sustainable transport raised in selected references. (including References a–p to Table 1); note that the allocation of issues to dimensions is somewhat arbitrary

	Environmental	*Economic*	*Social*
Development (Present generation)	Healthy air quality Acceptable noise Limited pollution/Waste Visual quality/liveability	Mobility and access Travel time/congestion Travel costs and prices	Safety Equity in mobility/ access
Sustainability (Future generations)	Climate stability Protecting ecosystems/ biodiversity Land conservation Resource conservation	Transport reinvestments Transport innovations Economic viability	Intergenerational equity in mobility Community cohesion

Chains',[37] and even 'The Sustainable Car'[38] abound. Such examples raise additional problems since it can be argued that sustainability is a system level condition, suggesting that the full context of interactions of any entity must be taken into account before its sustainability or the contrary can be claimed.[39] This means that, for instance, the number and actual use made of so-called 'sustainable' cars must be considered with respect to general system limits. In any case sustainable transport is most often conceived within specific political boundaries defined by either a state or a metropolitan region. Such boundaries may again be questioned, for instance, with respect to upstream or downstream environmental impacts from transport occurring outside the particular region, or by reference to problems such as transiting traffic.[40]

Operationalization

Operationalization concerns what to do to make the concept of sustainable transport manipulable or measurable and how to provide for interpretation and decision on that basis. In the following we consider indicators and address briefly their roles in this respect.

Indicators. An indicator may be defined in technical terms as a variable representing an operational attribute of a system.[41] Indicators are selected and

Zusammenarbeit (GTZ), Eschborn, 2001.

[37] J. Cooper, I. Black and M. Peters, Creating the sustainable supply chain: modelling the key relationships, in D. Banister, (ed.), *Transport Policy and the Environment*, Spon, London and New York, 1998, 176–203.

[38] N. S. Ermolaeva, K. G. Kaveline and J. L. Spoormaker. Materials selection combined with optimal structural design: concept and some results, *Mater. Design*, 2002, **23**, 459–470.

[39] E. Tengström, *Towards Environmental Sustainability? A Comparative study of Danish, Dutch and Swedish Transport Policies in a European Context*, Ashgate, Aldershot, 1999.

[40] This may be the reasoning behind the claim: 'There can be no understanding of sustainability at any level other than global', J. Whitelegg, *Transport for a Sustainable Future. The Case for Europe*, Belhaven Press, London, 1993, p. 11.

[41] G. C. Gallopin, Indicators and their use: information for decision-making, in B. Moldan and S. Billharz, (eds.), *Sustainability Indicators. Report on the Project on Indicators of Sustainable Development*, Wiley, Chichester, 1997, 13–27. Another reference suggests that definitions of

constructed from underlying data to condense complex information into a simplified form, providing a significant message about the system of interest. Indicators are used in many types of communication, from scientific analysis to every-day interaction. Indicators are widely used in most areas of policy analysis and policy making, not least environmental policy.

Different types of indicators convey different types of messages. The European Environment Agency distinguishes between the following types:[42]

1. Descriptive indicators, measuring state or trend in some entity or area. For example, the emissions of carbon dioxide from transport modes in Europe.
2. Performance indicators, comparing state or trend with a standard, norm or benchmark. For example, the number of dwellings affected by noise compared with a target number.
3. Efficiency indicators (ratios, or combining related descriptive trends). For example, average fuel efficiency of new vehicle registrations.
4. Policy effectiveness indicators (the role of policy in observed changes). For instance, the effect of Emission limit legislation on actual emissions of NO_x from motor vehicles.
5. Indices aggregating several indicators into one message. For instance, combining several air pollutants into indices of acidification, ozone formation, or global warming.

Often there is a wish to apply complex types of indicators (type 2–5) to represent some problem, but in practice even basic descriptive indicators can be difficult to establish due to a lack of reliable data series. Data quality is a very important concern in operationalization, since unreliable data can mean that communication is distorted rather than facilitated. For some transport indicator systems like ones used by the European Environment Agency,[43] and the United States Department of Transportation,[44] systematic assessments of the quality and reliability of the underlying data are made available.

Other important operational characteristics of indicators include:[45]

- Provision of a representative picture
- Simplicity, reduction of complexity
- Responsiveness to changes

indicators vary according to application but always . . . 'refer to a variable that is directly associated with a latent variable such that differences in the values of the latent variable mirror differences in the values of the indicator.' (K. A. Bollen, Indicator: methodology, in P.B. Baltes and N. J. Smelser (eds.) *The International Encyclopaedia of the Social and Behavioral Sciences*, Elsevier Science Ltd, Oxford, 2001, 7282–7287.

[42] P. Bosch, The European Environment Agency focuses on EU-policy in its approach to sustainable development indicators, *Statistical J. UN ECE*, 2002, **19**, 5-18.

[43] EEA, Paving the way for EU enlargement, *Indicators of Transport and Environment Integration TERM* 2002, Environmental issue report No 32, European Environment Agency, Copenhagen, 2000.

[44] Bureau of Transportation Statistics, *Source & Accuracy Compendium for Performance Measures in the DOT 2001 Performance Plan and 1999 Report*, Bureau of Transportation Statistics, U.S. Department of Transportation, Washington DC, 2000.

[45] OECD, *Environmental Indicators, OECD Core Set of Indicators for Environmental Performance Reviews*, Environment Monographs No 83. OECD/GD (93)179, Organisation for Economic Co-operation and Development, Paris, 1993.

- Theoretical founding in technical and scientific terms
- Adherence to international standards and international consensus about its validity
- Updating at regular intervals in accordance with reliable procedures.

Sustainability Interpretation. Not all types of indicators are equally relevant for reporting in a sustainable development context. If 'development' is understood as improvement in some aspect of immediate human interest (welfare, quality-of-life, environmental nuisance *etc.*), the trend may be derived from descriptive indicators, whereas 'sustainability' indicators should aim to '... *reflect the reproducibility of the way a given society utilises its environment*'.[46] In a broader sense sustainability indicators may be expected to describe critical systemic properties or trends.[47] This means that basic descriptive indicators in themselves often will provide only limited guidance towards sustainability, whereas performance indicators or indices—if devised correctly and substantiated with reliable data—would be more to the point. Ideally, if one aggregate index of sustainable transport was available a positive or negative trend in this index would suffice. In practice, a meaningful index is difficult to apply because weighting factors to aggregate items such as, say, transport CO_2 emissions, resource depletion and accessibility can be challenged. How to interpret contradicting trends in a sustainability context may prove difficult. Obviously, the usefulness of particular types of indicators depends on how sustainable transport is approached in a specific context. The full range of indicator types may therefore be useful. The interpretation of the results will nevertheless still be critical.

Confidence. A final matter of operational importance concerns the trust and confidence held in indicators by the presumed users and stakeholders. While this is dependent on relevance, reliability, and scientific soundness as addressed above it is also often referred to as a matter of involvement and participation, in particular in a highly normative context such as 'sustainability' measurement.[48]

Table 4 summarizes two sets of guidelines for the development of sustainability indicators. The sources are in wide agreement about key elements to consider, differing mainly in the level of detail in addressing each step.[49] However, none of the guidelines address the utilization phase of sustainability indicators, which we shall turn to now

[46] H. Opschoor and L. Reijnders, Towards sustainable development indicators, in H. Kuik, and O. Verbruggen, (eds.), *In Search of Indicators of Sustainable Development*, Kluwer, Dordrecht, 1991, p. 7.

[47] Wixey and Lake note: 'In practice, much of what passes for policy on "sustainable development" has a much narrower remit. The chief focus of much policy claiming to be "sustainable" is on issues more usefully placed under the category of "quality of life". Clearly there is a distinct overlap between "sustainable development" and "quality of life" agendas, but it would be foolhardy to equate them. Sustainable development goes beyond a shallow environmental approach or short term concern for living standards.' (S. Wixey, and S. Lake, Transport policy in the EU: a strategy for sustainable development? *World Transport Policy and Practice*, 1998, **4**, No. 2).

[48] J. E. Innes and D. E. Booher, Indicators for sustainable communities: A strategy building on complexity theory and distributed intelligence, *Planning Theory and Pract.*, 2000, **1**(2), 173–186.

[49] The two references also diverge in the view of the environmental dimension—if it is to be considered as a fundamental pillar of sustainability in itself or if rather the integration of three dimensions is the basic notion.

Table 4 Key steps and elements in the design of sustainability indicators

Stage/element	*(1)*[a]	*(2)*[b]
Conceptualization	Define sustainability goals	Consensus on Principles
Sustainability concerns to be considered	• Integrating dimensions • Forward-looking • Distributional aspects • Participatory input	• Ecological system integrity • Futurity • Social equity (dirtibution) • Participation
	Scoping	Identify issues of concern
Operationalization	Choice of indicator framework Define indicator selection criteria	
	Identify potential indicators	Construct indicators Augment Quality of life indicators with reference to sustainability principles Modify to account for boundary difficulties Supplement with uncertainty indicators
	Evaluate and select final set Collect data and analyse results Prepare and present report	
	Assess indicator performance	Evaluate indicators with respect to objectives and characteristics

[a] V. W. Maclaren, Urban Sustainability Reporting. *J. Am. Plann. Assoc.*, 1996, **62**(2).
[b] G. Mitchell, A. May and A. McDonald, PICABUE: A methodological framework for the development of indicators of sustainable development, *Int. J. Sustain Develop. World Ecol.*, 1995, **2**, 104–23.

Utilization

Users, Uses, and Numbers. Even the best indicators are of little value if they are not used. Many functions and uses of indicators are reported in the literature. Initially, a distinction is often made between broad user groups: scientists, analysts, decision-makers and the general public. According to general understanding these users have different information needs and handling capacities. While scientists are trained to manipulate large amounts of data, top decision-makers and the general public are expected to prefer selected information on key issues of importance to avoid 'information overload'. In the terminology of indicators this is often envisaged as a layered pyramid (Figure 2) founded on data that provide input to a system of indicators to assist in planning and management, and from which a set of 'headline' indicators can be chosen (an example of the latter is given in Table 5).[50]

Limiting the number of indicators can serve to reduce the perceived complexity

[50] The simplification in this model is evident. Obviously scientists and analysts may also need to draw on aggregate indicators or use indices, while decision makers may be highly concerned with quite specific data.

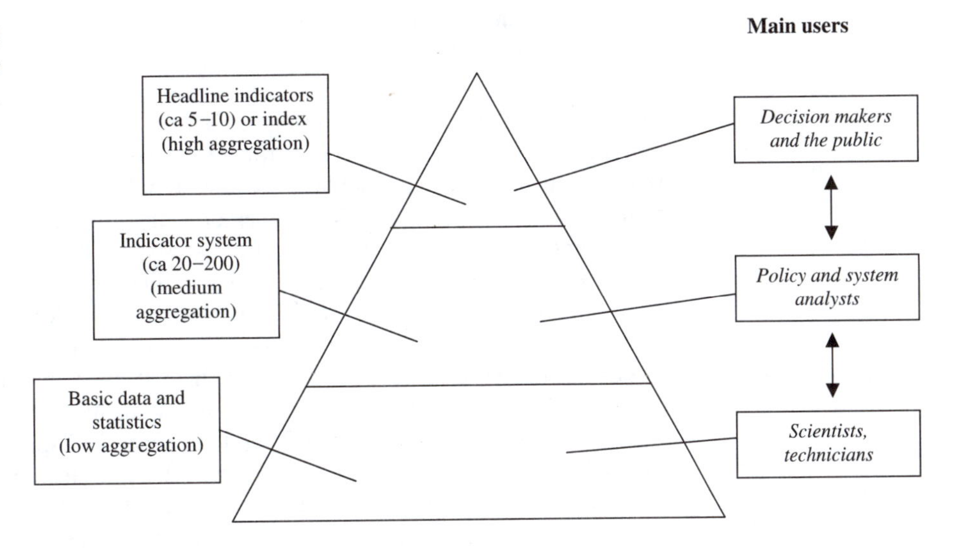

Figure 2 Simplified concept of 'Indicator pyramid'.

Table 5 Example of 'headline indicators' used by the EU. (Council of the European Union, *Environment-related Headline Indicators for Sustainable Development with a View to Monitoring Progress in the Implementation of the EU Sustainable Development Strategy — Council Conclusions*, Brussels, 28 November 2001)

Environment-related headline indicators
Combating climate change
• Greenhouse gases emissions, in absolute terms (related to Kyoto target)
• Share of renewables in electricity consumption
Ensuring sustainable transport
• Volume of transport *vs* GDP (passengers – km, freight in tonne – km)
• Modal split of transport (passengers – km, freight in tonne – km)
Addressing threats to public health
• Urban population exposure to air pollution
Managing natural resources more responsibly
• Municipal waste collected, landfilled and incinerated, in kg per inhabitant
General economic background
• Energy intensity of the economy (energy consumption/GDP)

of a particular issue. Conversely, important information may be lost in the aggregation or selection. Rather than necessarily minimizing the amount of indicators the challenge is to find the appropriate number for a particular context.

Indicator Functions. Four broad indicator functions will be distinguished here: (1) information, (2) assessment/forecasting/backcasting, (3) evaluation/monitoring and (4) control. These functions can either be embedded in customized indicator frameworks, or one set of indicators can aim to support several functions.

Information-oriented indicator frameworks can help policy makers, stakeholders or the general public increase their understanding and awareness of, for instance, environmental issues. The indicators used are typically descriptive rather than performance or index types. No specific requirements to use the information will normally exist, but information frameworks may nevertheless be influential in helping to form a basic, common awareness of a problem. In some cases the process of defining simple indicators has helped in fostering consensus on a course of action in local sustainable transport planning.[51]

Forecasting and assessment are integral tasks in transport planning and policy preparation. In these contexts indicators may be employed to compare expected trends or the results of a particular project with policy objectives concerning, *e.g.*, traffic volumes or emissions. The types of indicators used can be descriptive, performance, index *etc.* In the context of assessment, 'sustainability' is in some cases considered as an overall objective while in others it refers only to environmental effects not covered by standard Cost–Benefit assessment procedures.[52] *Backcasting* is the reverse case, where objectives are defined and necessary changes to achieve them assessed in an iterative process. This approach has been suggested as particularly relevant in sustainable transport analysis, considering the potential need for 'trend breaches'.[53] The utilization made of assessment and fore/backcasting results will depend entirely on the particular circumstances but often there is a presumption that rational decision-making would at least take into account results indicated in well-performed studies.

Use of indicators in *evaluation and monitoring* is frequent in transport policy, especially to compare developments in transport systems or policies over time and space. Evaluation and monitoring will often rely on performance indicators, for instance by gauging the present situation or trend by a policy objective. Evaluation is typically a one-time event while monitoring provides repeated feedbacks to decision making. Evaluation and monitoring procedures are often infused with a strong pretence of policy utilization, but in practice direct instrumental use is not always found to take place.[54] A particular application is *benchmarking*, where the best performer in a certain area (*e.g.* public transport punctuality) is identified and used as a basis for comparison and possibly transfers of effective practices.[55] This usage further emphasizes the need for *comparability*.

Control frameworks employ indicators to steer actions towards desired objectives. This function is entirely dependent on the use of performance indicators. A control function exists if the indicators are systematically used to direct or-

[51] European Commission, Integration of environment into transport policy—from strategies to good practice. *Highlights from the Conference on Good Practice in Integration of Environment into Transport Policy, 10–11 October 2002, Brussels, Belgium*, A Sourcebook, European Commission, Directorate-General Environment, Brussels, September 2003.

[52] L. Giorgi and A. Tandon. The theory and practice of evaluation, *ICCR Working Paper 407*, The Interdisciplinary Centre for Comparative Research in the Social Sciences (ICCR) Vienna, June 2000.

[53] D. Banister, D. Stead, P. Steen, J. Akerman, K. Dreborg, P. Nijkamp and R. Schleisher-Tappeser, *European Transport Policy and Sustainable Mobility*. Spon Press, London and New York, 2000.

[54] See for instance: L. Shulha, J. Cousins, and M. Bradley, Evaluation use: theory, research, and practice since 1986. *Evaluation Practice*, 1997, **18**(3), 195–208 and E. Vedung, *Public Policy and Program Evaluation*, Transaction Publishers, New Brunswick, 1997; also H. Gudmundsson, The use of environmental indicators—Learning from evaluation research, *J. Transdiscipl. Environ. Stud.*, 2003, Vol. 2, No. 2.

[55] ECMT, Transport benchmarking: methodologies, applications and data needs'. *Proceedings of the Paris Conference, November 1999*, European Conference of Ministers of Transport, OECD Studies, September 2000, Vol. 1, No. 9.

Table 6 Perspectives of different indicator frameworks

Backwards	*Forwards*
Information	
	Assessment/forecasting/backcasting
Evaluation/monitoring	
Control	

ganizations or entities to adapt their activities according to measured or stipulated results, using penalties, rewards, budget allocations or similar incentives. This type of framework spans future and past by conditioning future actions on past results (Tablc 6). Control frameworks are widely used in managing transport contracts and outsourced services, where penalties may for instance be routinely issued if punctuality or other service targets are not met. Another example is the performance-based budgeting approach adopted by the United States government. In this context the US Department of Transportation is obliged by law to produce annual performance plans and reports, which are tied into the national budget process.[56]

Sustainable transport indicators could in principle be incorporated into all of the above frameworks. Their use and impact may, however, be very different, depending on the particular kind of utilization framework and practice. Considering the challenges involved in making sustainable transport operational one may expect that strongly control oriented frameworks would be difficult to apply for this purpose, since the exercise of control tends to depend on solid data and transparent concepts. Conversely, the rigor of a control framework could push operationalization forward, albeit it may entail a narrowing of perspective.

Summing up

To make sustainable transport into an operational concept that is measurable by indicators and useful for policy action is a complex task. The process may be hampered by difficulties in steps such as definition, boundary drawing, measurement, interpretation, confidence and utilization. Nevertheless many suggestions for performance indicators for sustainable transport have been made. From the analysis in this section we may infer that such indicators from an ideal point of view could be challenged to address the following points:

- Concern for both present (development) and future (sustainability) generations
- Consideration of all dimensions (economic, social, environmental, institutional)
- Identification of key transport contributions and shares of overall problems
- Considerations of transport system boundary (and induced effects elsewhere)
- Inclusion of sustainability criteria or targets to interpret performance
- Insurance of data quality, reproducibility, *etc.*
- Participation of stakeholders in indicator development
- Adoption of an appropriate number of indicators
- Catering to maximum utilization and impact

[56] For instance: US DOT, *1999 Performance Report 2001 Performance Plan*, US Department of Transportation, Washington, DC 2000.

3 Sustainable Transport Indicator Sets

Overview

Over the last decade or so several indicator sets addressing sustainable transport have been defined in various contexts. Academics or consultants are the authors of most of the sets explicitly referring to sustainable transport, while the officially adopted indicator systems tend to have a more pragmatic focus on transport and/or environmental policy issues. Table 7 provides an overview of a number of sets and systems of both types. In the following, we briefly review four of them, highlighting differences in approach with respect to the context and some of the criteria mentioned above (sustainability dimensions, indicator types, and interpretation). Particular features of interest from each will be highlighted. The aim is neither to make a comprehensive assessment of the systems nor to compare them directly. The purpose is rather to give an impression of various ways to tackle the problems involved in measuring and reporting on sustainable transport in different contexts.

(1) Lyon Study

Background and Purpose. A French research group has defined a set of indicators to monitor sustainable transport at the urban level.[57] The indicators have been applied to the case of Lyon, France (Table 8). The further aim is to extend the study to other cities in France and Europe. The analysis exploits a detailed passenger travel survey combining this information with various environmental and other data. The purpose is to provide analytical information to policy makers and the public. The indicators (types and number) appear to have been selected by the research group alone.

Sustainability Dimensions, Indicator Types and Other Features. The 'Lyon case' indicators explicitly aim to cover three dimensions of sustainability: environmental, social and economic. General mobility measures are also included (see Table 8). Most of the indicators measure aspects related to quality of life for the present generation. More long-term issues are also represented, including indicators for CO_2 emissions, energy consumption and land take. All of the indicators are descriptive rather than performance. Environmental efficiency by mode is also calculated. A strong feature of the study is the combination of indicators from different dimensions. This enables, for instance, a spatial breakdown of emissions, both in terms of where the more polluting trips are generated (by socio-economic groups) and where the emissions occur (emission densities). An interesting indicator measures space occupancy of transport mode in m^2 hours. Calculating the public space occupied by vehicles both while driving and parked over time suggests that private motor cars account for 96% or more than 40 times that of public transport, even though they account for only about 4 times the passenger transport.

[57] J.-P. Nicolas, P. Pocheta, and H. Poimboeuf. Towards sustainable mobility indicators: application to the Lyons conurbation, *Transport Policy*, 2003, **10**, 197–208.

Interpretation. The study is descriptive and the authors abstain from engaging in normative or performance-based considerations.[58] Therefore, the interpretation of trends in terms of sustainability is entirely open. Planned appliations to other French cities will enable comparisons that may reveal further implications. However, the authors suggest that the main relevance of their approach for sustainable transport is to consider the three dimensions separately, rather than to attempt any aggregation. In addition, they suggest an interpretation inspired by a paretian[59] approach according to which increased sustainability would mean any improvement in one dimension without deterioration in any other. This is equivalent to the notion of strong sustainability introduced in Section 2.

(2) Sustainable Transport Index — USA

Background and Purpose. In a US study[60] five key threats to the sustainability of transport developments are identified as: petroleum scarcity; impact of emissions on local air quality and human health; the impact of emissions on the atmosphere; excessive number of injuries and fatalities; and high levels of congestion. The purpose of the study is to derive an index of transport sustainability reflecting these and other factors based on already existing data. The index is used to compare US states in terms of their sustainable transport performance (with possible extensions to other countries, cities, *etc.* if comparable data are made available). The aim is to facilitate the inclusion of sustainability concerns in transport analysis and policy debates. The study draws inspiration from previous academic research as well as policy studies, but the proposed index is derived and used independently by the author.

Sustainability Dimensions, Indicator Types and Other Features. A qualitative judgement forms the basis for identifying the issues that represent the major possible threats to transport sustainability. They address the environmental and economic dimensions, while issues of social equity are explicitly excluded.[61] In addition to the threats also a number of presumed positive measurers of transport sustainability are identified, including public transport ridership and the number of alternative fuelled vehicles. The indicators used to construct the index are summarized in Table 9. All indicators are descriptive and directly drawn from existing general statistics covering the 50 US States. Some indicators have a quite indirect relation to the issue of interest (in particular congestion measured as classes of road traffic flows). The data material is used to derive an aggregate index of transport sustainability using Principal Components Analysis. The method reveals an underlying or latent variable assumed to be most representative of the data, in this case indicating a measure of transport sustainability.

The results show strong variation in the sustainability index between the US states. Since the data represent absolute values it is obviously the largest states (Texas, California) that come out as least sustainable. Interestingly, this does not

[58] Reference 57, p. 207.

[59] After the Italian Economist Vilfredo Pareto (1848–1923).

[60] W. R. Black, *Toward a Measure of Transport Sustainability*, Transportation Research Board Meeting, 2000, Conference Pre-prints, Transportation Research Board, Washington, D.C.

[61] Reference 60, p. 2.

Table 7 Overview of selected Sustainable Transport indicator sets

Ref.	*Status*	*Focus*	*Level*	*Purpose*	*No. of Indic.*	*Dimensions (main emphasis in italics)*
a	Official	Transport system + policy	International (Europe)	Monitoring	≈35	*Environmental*, economic, social, institutional, transport
b	Official	Transport system + policy	International (OECD)	Monitoring	≈14	*Environmental*, economic, social
c	Expert	Transport system	International (Europe)	Backcasting	≈7	*Environmental*, economic, social
d	Official	Transport system	National (USA)	Evaluation	≈166	*Environmental*
e	Official	Transport policy	National (Canada)	Monitoring	≈80	*Environmental*, economic, institutional
f	Expert	Transport system	National (Canada)	Monitoring	≈14	*Environmental*, economic, social, transport
g	Expert	Transport system	National (USA)	Monitoring	≈14	Environmental, social, economic, transport
h	Expert	Transport system	National (Germany)	Forecasting	≈7	*Environmental*
i	Expert	Transport system	Sub-national (Hong Kong)	Monitoring	≈10	Environmental, social, economic, transport
j	Expert	Transport system + policy	Urban (General)	Evaluation	≈55	*Environmental*, economic, social, institutional, transport
k	Expert	Transport system	Urban (Lyon)	Monitoring	≈22	*Environmental*, economic, social, transport
l	Expert	Transport system	Urban (Siena)	Evaluation	≈14	*Environmental*
m	Expert	Transport System	Corridor (IS10 Texas)	Assessment	≈10	*Environmental*, economic, social, transport

[a] EEA, *Paving the way for EU Enlargement. Indicators of Transport and Environment Integration TERM 2002*, Environmental issue report No 32, European Environment Agency, Copenhagen, 2002.
[b] OECD, *Indicators for the Integration of Environmental Concerns into Transport Policies, ENV/EPOC/SE(98)1/FINAL* Organisation for Economic Co-operation and Development, Paris, 1999.
[c] D. Banister, D. Stead, P. Steen, J. Äkerman, K. Dreborg, P. Nijkamp and R. Schleisher-Tappeser, *European Transport Policy and Sustainable Mobility*. Spon Press, London and New York, 2000.
[d] US EPA, *Indicators of the Environmental Impacts of Transportation*. Updated Second Edition, United States Environmental Protection Agency, Washington DC, 1999.
[e] Transport Canada 2001–2002 Sustainable Development, *Strategy Progress Report*, Ottawa 2002 URL: http://www.tc.gc.ca/programs/environment/sd/sds0102/menu.htm.
[f] R. Gilbert, N. Irwin, B. Hollingworth, and P. Blais, Sustainable Transportation Performance Indicators (STPI). *Project Report On Phase 3*, The Centre for Sustainable Transportation, Toronto, 2002.
[g] W. R. Black, Toward a measure of transport sustainability, *Transportation Research Board Meeting, 2000, Conference Preprints*, Transportation Research Board, Washington, D.C., 2000.
[h] J. Borken, Indicators for sustainable mobility – a policy oriented approach, *Proceedings: 1st International Scientific Symposium 'Transport & Environment'*, Avignon/France 19–2. June 2003. R. Joumard (ed.), *Actes INRETS No. 93*, Arceuil, 2003.
[i] W.-T. Hung, Indicators for sustainable transport policy. Paper presented at: *Better air Quality Motor Vehicle Control & Technology Workshop 18–20th September 2000*, Hong Kong, 2000.
[j] L. L. Ricci, Monitoring progress towards sustainable urban mobility, *Evaluation of Five Car Free Cities Experiences*, EUR 19748 EN, Institute for Prospective Technological Studies, Seville, December 2000.
[k] J.-P. Nicolas, P. Pochet, and H. Poimboeuf, Towards sustainable mobility indicators: application to the Lyons conurbation, *Transport Policy*, 2003, **10**, 197–208.
[l] M. Federici, S. Ulgiati, D. Verdesca and R. Basosi, Efficiency and sustainability indicators for passenger and commodities transportation systems, The case of Siena, Italy, *Ecol. Indicators*, 2003, **3**, 155–169.
[m] J. Zietsman and L. R. Rilett, *Sustainable Transportation: Conceptualization and Performance Measures*. Report No. SWUTC/02/167403-1, Texas Transportation Institute, College Station, Texas, 2002.

Table 8 Indicators of sustainable transport in the Lyon study (Reference 57)

Dimension of sustainability	*Indicators*
Mobility	
Service provided	Daily number of trips Structure of trip purposes Daily average time budget
Organization of urban mobility	Modal split Daily average distance travelled Average speed (global and per person)
Economic	
Cost for the community	Annual costs chargeable to residents of the conurbation, due to their mobility in this zone
Expenditures of the participants involved	Households: Annual average expenditures for urban mobility (per person) Companies: Costs of employee parking Subsidies to employees (company cars *etc.*), possible local taxes Public authorities: Annual expenditures for investments and operates
Social	Proportion of households owning 0, 1 or more cars Distance travelled Expenditures for urban mobility: Amounts for private/public transport; for fixed/variable cost of car; share of the average income of households
Environmental	
Air pollution–Global	Annual energy consumption and CO_2 emissions (total and per resident)
Air pollution–Local	Levels of CO, NOX, HC and particles (in g m^{-2}, total and per resident)
Space consumption	Daily individual consumption of public space involved in travelling and parking (in m^2 h) Space taken up by transport infrastructures
Other items[a]	Noise intensity levels Risk of accident

[a] Not included due to lack of data.

alter after correction for population size, while the same operation does change significantly the sequence in the leading end of the table, where relatively undeveloped states such as Wyoming, Montana and Vermont score the highest. The analysis also shows that the 'positive' indicators (public transport ridership, alternative vehicles) influence the index very little. The best 'proxy' for the latent index variable is vehicle miles travelled.

Interpretation. The obvious interpretation of the index is that some US states have a better transport situation than others do. To the extent the index reflects

Table 9 Issues and indicators in the Sustainable Transport Index (Reference 60)

Issue	*Indicator*
Global atmospheric pollution	Carbon dioxide emissions
Local air pollution	Carbon monoxide emissions
	Nitrous oxides emissions
	Volatile organic compound emissions
Dependence on petroleum fuels	Gasoline sales
	Number of motor vehicles
	Vehicle miles of travel
Accidents	Fatalities
	Injuries
Congestion	Urban population
	Miles of road with with 40 000 annual daily traffic
Use of mass public transit	Transit riders
Use of gasohol	Gasohol sales
Use of alternative fueled vehicles	Alternate fueled vehicles

important conditions for sustaining transport systems it may be revealing. The index does not suggest any thresholds limiting the continuation of current trends, nor does it suggest the importance of the distance in ranking between best and worst performing states. As noted by the author, another approach using the same data could be to assign explicit weights to each of the indicators. While this would not provide a more objective assessment of sustainability it could perhaps cater to a more participatory use of indicators. As also noted, the method could as well be used to 'backcast' the required increase, *e.g.* in passenger ridership, to advance a particular state in the ranking.

(3) 'TERM'—European Union

Background and Purpose. The Transport and Environment Reporting Mechanism (TERM) is an official monitoring framework and system set up by the European Environment Agency (EEA) in collaboration with the EU Commission and EUROSTAT. The mechanism consists of indicator reports based on an extensive database, with three (annual) reports so far.[62] TERM reports have covered around 35–40 different indicators (Table 10) for a wide range of transport and environment trends in EU Member States as well as accession countries. The main purpose is to support the political process of integrating environmental concerns into transport policy. TERM can be considered as influential since it is a pioneering system in EEAs sectoral monitoring efforts and since it draws on a large network of institutions throughout Europe.[42] TERM serves information, assessment and monitoring functions but is not linked to a control system.

[62] EEA, Paving the way for EU enlargement, *Indicators of Transport and Environment Integration TERM 2002*, Environmental issue report No 32, European Environment Agency, Copenhagen, 2000; EEA, *TERM 2001, Indicators Tracking Transport and Environment Integration in the European Union*, European Environment Agency, Copenhagen, 2001; EEA, Are we moving in the right direction? *Indicators on Transport and Environment Integration in the EU. TERM 2000*, Environmental issues series No 12, European Environment Agency, Copenhagen, 2000.

Table 10 Indicators in the TERM system. Note that not all indicators are represented by actual data (EEA. Paving the way for EU enlargement. TERM 2002. Environmental issue report No 32. European Environment Agency, Copenhagen, 2000

1 Environmental consequences of transport	Transport final energy consumption and primary energy consumption, and share in total by mode and by fuel
	Transport emissions of greenhouse gases (CO_2 and N_2O) by mode
	Transport emissions of air pollutants (NO_x, NMVOCs, PM_{10}, SO_x, total ozone precursors) by mode
	Population exposed to exceedances of EU air quality standards for PM_{10}, NO_2, benzene, ozone, lead and CO
	% of population exposed to and annoyed by traffic noise, by noise category and by mode
	Fragmentation of ecosystems and habitats
	Proximity of transport infrastructure to designated areas
	Land take by transport infrastructure by mode
	Number of transport accidents, fatalities, injured, and polluting accidents (land, air and maritime)
	Illegal discharges of oil by ships at sea
	Accidental discharges of oil by ships at sea
	Waste from road vehicle (end-of-life vehicles)
	Waste from road vehicles (number and treatment of used tyres)
2 Transport demand and intensity	Passenger transport (by mode and purpose)
	Freight transport (by mode and group of goods)
3 Spatial planning and accessibility	Access to basic services: average passenger journey time and length per mode, purpose (commuting, shopping, leisure) and location (urban/rural)
	Regional access to markets: the ease (time and money) of reaching economically important assets (e.g. consumers, jobs), by various modes (road, rail, aviation)
	Access to transport services
4 Supply of transport infrastructure and services	Capacity of transport infrastructure networks, by mode and by type of infrastructure (motorway, national road, municipal road, etc.)
	Investments in transport infrastructure/capita and by mode
5 Transport costs and prices	Real change in passenger transport price by mode
	Fuel prices and taxes
	Total amount of external costs by transport mode (freight and passenger); average external cost per passenger-km and tonne-km by transport mode
	Implementation of internalisation instruments i.e. economic policy tools with a direct link with the marginal external costs of the use of different transport modes
	Subsidies
	Expenditure on personal mobility per person by income group

6 Technology and utilisation efficiency	Overall energy efficiency for passenger and freight transport (per passenger-km and per tonne-km and by mode)
	Emissions per passenger-km and emissions per tonne-km for CO_2, Nox, NMVOCs, PM_{10}, Sox by mode
	Occupancy rates of passenger vehicles
	Load factors for freight transport
	Uptake of cleaner fuels (unleaded petrol, electric, alternative fuels) and numbers of alternative-fuelled vehicles
	Size of the vehicle fleet
	Average age of the vehicle fleet
	Proportion of vehicle fleet meeting certain air and noise emission standards (by mode)
7 Management integration	Number of Member States that have implemented an integrated transport strategy
	Number of Member States with a formalised cooperation between the transport, environment and spatial planning ministries
	Number of Member States with national transport and environment monitoring systems
	Uptake of strategic environmental assessment in the transport sector
	Public awareness and behaviour
	Uptake of environmental management systems by transport companies

Sustainability Dimensions, Indicator Types and Other Features. TERM is not explicitly aiming to monitor sustainability. Nevertheless, it covers a wide range of the environmental issues in sustainable transport of relevance to present and future generations, including emissions, fragmentation of land, noise, waste, and oil spills. It also describes some economic factors. TERMs main focus is to monitor developments in key areas of policy intervention, such as improvements in technology, investments in infrastructure, changes in prices and taxes, changes in the spilt between modes of transport and changes in the institutional frameworks of decision making. The latter is a special feature of TERM not found in many other frameworks: Monitoring the way EU Member State governments integrate environmental concerns in their organizations and procedures, for instance by undertaking Strategic Environmental Assessments. (Topic 7, 'management integration' in Table 10). Most of the indicators are descriptive, but there are also occurrences of eco-efficiency and policy effectiveness indicators. There are no quantitative performance indicators, due to a lack of quantitative objectives for transport and environment at the European level. There are a few qualitative performance indicators, regarding environmental management integration issues (topic 7), of the dichotomous type ('yes/no' to the presence/absence of certain measures).

Interpretation. The direction of change as favourable or unfavourable is clearly signified by use of the 'Smiley' symbol for each indicator. The 2001 report conclude that . . . '[o]*verall, the report shows that transport is becoming less and not more environmentally sustainable, and integration efforts have to be redoubled.*'[63] Since the indicator system does not (claim to) provide a criterion or aggregate measure for sustainable transport the basis for this conclusion is not quite clear. One interpretation could simply be that the majority of the indicators are in the negative in all three reports so far (for instance increase in emissions of CO_2, increasing pressures on nature, *etc.*) while only a few are positive (*e.g.* emissions of acidifying substances). Another interpretation could be that the EU has stated 'decoupling of transport growth from economic growth', 'stabilization of modal split' and 'environmental integration' as key objectives of its sustainable transport policy. Since the indicators for those objectives are mostly negative, one could infer that transport in EU is moving away from policy commitments to sustainability.

(4) Transport Canada

Background and Purpose. As a ministry of the Canadian Government, Transport Canada is obliged by law to produce a Sustainable Development Strategy (SDS), and to monitor progress in its implementation. Its first SDS was adopted in 1997 and the second revised one in 2001. The SDS is structured around a set of seven so-called Challenges (Table 11), broken down into 29 Commitments, and approximately 80 targets and performance indicators. Most indicators refer to progress in actions to be taken by Transport Canada to fulfil the strategy (*i.e.* take steps to implement a certain policy measure). A first review of progress was made in 2003.[64] This review was fed into Transport Canada's so-called Departmental Performance Report,[65] which is a part of the preparation process for the political adoption of the national budget. The system can then be described as a monitoring system, linked to a control system.

Sustainability Dimensions, Indicator Types and Other Features. The strategy has a focus on the environmental dimension of sustainability. It concerns, in particular, institutional and policy aspects of the environment rather than physical environmental results.Most of the indicators are of the performance type. However, performance is not measured against quantitative sustainability targets but against the mostly qualitative policy commitments. To illustrate the approach a few extracts from the monitoring report are shown in Table 12.

The commitments are mostly of a short-term character and rather detailed. In this way the SDS monitoring system enables a quite specific assessment of what the ministry is doing or not doing. It can thereby support the notion of governmental accountability towards the public, in this case concerning the

[63] EEA, TERM 2001, *Indicators Tracking Transport and Environment Integration in the European Union*, European Environment Agency, Copenhagen, 2001, p 3.

[64] Transport Canada 2001–2002 Sustainable Development, *Strategy Progress Report*, Ottawa 2002. URL: http://www.tc.gc.ca/programs/environment/sd/sds0102/menu.htm.

[65] Transport Canada, *Departmental Performance Report. For the period ending March 31, 2003*, Minister of Transport, Ottawa, 2003.

Table 11 Challenges for sustainable transportation (Reference 65)

Strategic challenges for sustainable transportation in Canada
Improving education and awareness of sustainable transportation
Developing tools for better decisions
Promoting adoption of sustainable transportation technology
Improving environmental management for Transport Canada operations and lands
Reducing air emissions
Reducing pollution of water
Promoting efficient transportation

extent to which policy efforts are being made to promote sustainability of transport. Beside the monitoring itself, the SDS and monitoring system is also reviewed by Internal management Control and by an external auditor, the Canadian Commissioner for Sustainable Development. There is a further controlling influence through the indirect linkage to the political budget process (not assessed here).

Interpretation. The review of the SDS indicated that about 80% of the commitments and 70% of the targets were either on track or complete. However according to the internal review by Transport Canada itself the SDS monitoring system does not establish direct links from measures to transport and environmental outcomes. A cautious interpretation of the results so far may therefore be that sustainability is increasing in the institutional dimension while the implications for sustainable transport at system level are unclear. Other initiatives have been taken to develop 'system level' monitoring of Canadian sustainable transport, but the results have not been implemented at this point.[66]

4 Discussion and Conclusion

Sustainable transport remains a challenge conceptually, but perhaps even more so for practical policy. Indicators are among the tools that can help in conceptual clarification as well as guide policy analysis, deliberation and decision-making. Many attempts have been made to specify and apply indicators in this field. This chapter has shown that there is not one uniform approach and not one general application—the function of sustainable transport indicators will be highly dependent on specific context, and can serve different users with different priorities and concerns.

Nevertheless, there appears to be agreement on many of the topics considered as important for sustainable development and transport, and the issues for which indicators are primarily defined. Chief among these topics are transport system contributions to climate change, regional air pollution, impairment of urban air quality, depletion of oil and land resources, as well as traffic induced death and injuries. Emphasis is also put on the impact of transport systems on biodiversity,

[66] R. Gilbert, N. Irwin, B. Hollingworth and P. Blais, *Sustainable Transportation Performance Indicators (STPI)*, Project Report on Phase 3. The Center for Sustainable Transportation, Toronto, 31 December 2002.

Table 12 Extracts from Transport Canada (Reference 65)

Challenge 5 reducing air emissions				
Commitment 5.1 Transport Canada will continue to lead the transportation component of the federal action plan on climate change . . .				
Targets (. . .) (c) Initiate discussions with the freight transportation industry in 2001 to establish voluntary initiatives to improve the fuel efficiency of the freight system	*Complete*	*On-track*	*No action to date*	*Behind*
	X			
Commitment 5.2 Transport Canada will work with ICAO to develop new aircraft emissions standards and operational practices that address concerns about local air quality and global climate change, from 2000/2001 – 2003/2004				
Targets (. . .) (a) Develop new engine standards that include emissions limits for nitrogen oxides during climb and cruise modes of flight, beginning in 2000/2001	*Complete*	*On-track*	*No action to date*	*Behind*
				X

ecosystems, and general natural resource funds, but while some indicators in these areas exist, operationalization is often more difficult. Most references also include immediate social and economic issues such as transport costs, congestion, and accessibility into the equation, even though their implications for sustainability (as opposed to development) often are unclear. There appears to be few attempts to conceptualize, let alone measure, the functions of transport in the general sustenance of social and economic systems over the longer term, for instance in terms of 'critical' levels of transport infrastructure, investment rates or innovation capacities. The institutional dimension of sustainable transport is included in some indicator systems, but how changes in this dimension links to system changes also appears to be an area for further exploration.

There is little agreement how to measure sustainable transport more exactly, how to aggregate the available information and even how to interpret results in terms of transport sustainability, beyond a general notion of listing 'positive' *versus* 'negative' contributions. A transport system may be deemed sustainable in some respect (for instance in terms of impact on the ozone layer) and unsustainable in some others (for instance in terms of contributing to climate change). What would that imply? Proposing that the 'majority' of a particular range of indicators should be positive to merit sustainability is of course arbitrary and would in any case make indicator (de)selection highly critical. Proposing that unsustainability in one dimension can be compensated by sustainability in another rests on weak assumptions about substitution between various forms of resources and services, which cannot always be assumed but would have to be established empirically. Converseley, maintaining that the transport sector is sustainable only if all indicators in all dimensions are positive or neutral assumes that improvements in other sectors can never compensate for negative transport trends. This may hold only if some critical system limit was irreversibly transgressed solely because of, and attributable to, a particular transport effect. To identify such a situation would be difficult, to say the least, and this chapter did not reveal operational criteria and measures to detect it. The absence of any such evidence should not, however, lead to the assumption that transport trends are positively sustainable.

Notwithstanding such speculations, indicators should not be assumed to provide definite answers to complex problems. This would negate the very meaning of indicators: as guidance to navigate in complex territories where exact and full knowledge is not available, but where actions are necessary after all. At best, indicators can help to reduce the complexity of operation and communication. But does this mean that no particular line can be drawn between sustainable transport performance indicators and transport/environment indicators in general? Section 2 suggests some criteria that could help to police such a distinction. Adopting the full list would, however, most likely prohibit the identification of proper sustainable transport indicator systems and sets at all. As a minimum requirement the author would rather suggest that claims representing sustainable transport indicators were accompanied at least by (1) an explicit justification in terms of recognized notions of sustainability and development from the literature, (2) considerations over system boundaries and their implications, and (3) explicit reflections over which criteria on the list are included and which ones that are not considered in a particular application.

Policy Instruments for Achieving Sustainable Transport

DAVID BEGG AND DAVID GRAY

1 Introduction

Transport policy making is a multi-faceted discipline. It encompasses a number of fields, including economics, environmental sciences, engineering, traffic modelling, town and country planning, geography and sociology. However, in examining the recent history of transport policy making in the UK (in particular, progress since the publication of the Integrated Transport White Paper in 1998) it is perhaps more revealing to consider the political economy.

All Government policy is a trade-off between adhering to philosophical ideals and delivering policies that do not alienate the electorate of the time. Whereas the Labour Government of 1945 was able to introduce several radical reforms in the UK, the electorate of the early 21st Century is arguably more cynical and less supportive of social reform. This is particularly true of transport policy making, which is increasingly a compromise between sustainability and political practicality. Since it has emerged as a key political issue, wider concern over transport delivery has the potential to undermine the Government's prospects for future re-election. Moreover, while few academics, lobbyists or political commentators lose their jobs after advocating a controversial transport policy, the situation for many local politicians is rather more precarious.

In improving the UK's transport infrastructure, policy makers at local and national level must also conciliate a number of influential lobbies. These include environmentalists, the motoring lobby, transport operators, the media and other opinion formers. Consequently, the political realities of transport policy making are central to any discussion of instruments for delivering 'sustainable' transport.

In examining the relationship between sustainable ideals and political reality, this review will consider the progress of the sustainable transport agenda in the UK since the publication of the Integrated Transport White Paper. It will be argued that John Prescott's original integrated/sustainable agenda has been undermined by the need to appease a sceptical motoring public and a hostile

Issues in Environmental Science and Technology, No. 20
Transport and the Environment

press. As a result, a wholly sustainable transport policy has proved difficult to pursue, both locally and at national level.

It will be argued that the Government's need to deliver on transport without further alienating the motoring lobby could have significant implications for the environment. Nevertheless, it will be suggested that the introduction of national congestion charging would allow the Government to manage traffic levels, reduce congestion and continue the progress made in lowering levels of environmentally harmful emissions. The review will conclude with a discussion of the technological and political barriers that stand in the way of a national charging scheme.

Much of the analyses contained here deals with policy making in England. Transport policy is a devolved power in Scotland and Wales. Nevertheless, while there are certain policy differences between Westminster and the devolved administrations, many of the issues are relevant to the UK as a whole.

2 'Sustainable' Transport in the UK: the Transport White Paper

Concerns over the sustainability of the UK's transport system have emerged in response to increasing dependence on the car at the expense of other—more environmentally friendly—modes.

Reliance on the car has been driven by—and has in turn prompted—long-term social change. Over the last two or three decades, motoring costs have fallen in real terms, while long-term economic growth has increased the public's buying power. Car ownership is now affordable for a majority of households in the UK. There were 23.9 million cars registered in the UK in 2001 compared to 19.7 million in 1990. Similarly, the proportion of households with access to at least one car increased from 67% to 72% over the same period.[1]

Although increasing car dependence has been facilitated by motoring becoming more affordable (the cost of a car has decreased substantially in real terms, while modern-day vehicles also last longer and are more reliable), it is also bound up and interrelated with other social trends that have created a more dispersed society, one increasingly reliant on levels of individual mobility. Although many of the social changes underlying this rise in car-based mobility would have happened anyway, policies introduced by the previous Conservative Government hastened a trend towards growing dependence on the car at the expense of public transport and, increasingly, walking.[2] Planning, housing, retail and economic development policies fostered out and edge-of town development, and encouraged towns and cities to grow. Throughout the 1980s and 1990s, access to employment, shops and services became increasingly dependent on access to a car, while flexible working practices resulted in a workforce increasingly divorced from its place of employment.[3]

[1] Department for Transport, *Transport Statistics: Transport Trends 2002*, 2003, available http://www.transtat.dft.gov.uk/tables/2002/tt/index.htm (accessed 1st October 2003).

[2] M. Buchanan and G. Urquhart, 18 Years of Tory transport policy, *New Government—New Transport, Edinburgh, 10th-11th July 1997*. Transport Research Institute, Napier University, Edinburgh, 1997.

[3] D. Begg and D. Gray, Transport Policy and vehicle emission objectives in the UK: is the marriage

As a result, motorists make ever more and longer journeys. The average annual distance travelled per person by car rose from 3199 miles in 1975/76 to 5354 miles in 1999/ 2001.[1] In the 1990s alone, the distance travelled by car increased by 11% while the number of trips per person per year made by car rose by 3% over the same period. Symptomatically, average commuting distances have also increased. The Commission for Integrated Transport reports that commuters in the UK now travel longer distances than those in any other European countries,[4] while 72% of commuters outside London travel to work by car.[5]

Transport Policy and Sustainability

Successive Governments have had to deal with the impact of increasing traffic levels. The previous Conservative Government's initial approach to rising traffic levels was to create additional road space through an ambitious road building programme.[6] In the late 1980s and 1990s, however, mounting environmental awareness among the public was increasingly impacting on transport policy, and by 1996 pressure from environmentalists and the Treasury led to a U-turn on road building. Prior to their election defeat in 1997, the Conservatives were giving road charging serious consideration.

Concerns about the environment were also reflected in the introduction of the Fuel Duty Escalator (FDE), established by Chancellor Norman Lamont in 1992. In a bid to limit the growth of vehicle emissions, an annual increase in fuel duty was introduced, initially at 3%. The annual increase was subsequently raised to 5% by Chancellor Kenneth Clark in 1995. Clark robustly defended the increase, stating that '*anyone who is opposed to the Fuel Duty Escalator and is pro-environment is guilty of gross hypocrisy.*'[6]

In 1997, a Labour Government was elected, and the Department of the Environment, Transport and the Regions (DETR) was created under Deputy Prime Minister John Prescott. Recognizing the problems of increasing car dependence and the interrelationship between transport, planning, economic development, Prescott intended for DETR to take a more 'joined-up' approach to transport planning and the environment, and this philosophy was embodied in the Integrated Transport White Paper (ITWP), published in 1998.

With the aim of pursuing a more sustainable transport policy, the ITWP set out a fully integrated policy to tackle the two key priorities of congestion and pollution.[6] This philosophy was set out clearly in the Foreword:

'*the previous Government recognised that we could not go on as before*, building *more and more new roads to accommodate the growth in car traffic. With our new*

between transport and environment policy over? *Environ. Policy Sci.*, 2004, **7**(3), 155.

[4] Commission for Integrated Transport, *European Best Practice in Delivering Integrated Transport: Key Findings*, 2001, available http://www.cfit.gov.uk/research/ebp/key/index.htm (accessed 1st October 2003).

[5] RAC Foundation, *Motoring Towards 2050 — An Independent Inquiry*, (RAC Foundation, London, 2002.

[6] D. Gray and D. Begg, *Delivering Congestion Charging in the UK: what is required for its successful introduction?*, The Centre for Transport Policy: Policy Paper Series, No. 4, August 2001.

obligations to meet targets on climate change, the need for a new approach is urgent.'[7]

In seeking to integrate transport and environmental policy, the ITWP emphasized the inter-relationship between sustainable transport, development, planning and the environment. It also highlighted the need to reduce car use and encourage more sustainable modes (more walking, cycling and public transport), stressing the futility of simply building more roads.

The 10-Year Plan. The Government's route map for delivering its transport proposals was the *10-Year Plan for Transport*, published in 2000. Government spending in the UK is allocated through a four year spending cycle. Uniquely, Prescott was able to secure both additional resources and fiscal concessions from the Treasury, enabling DTLR to draw up a costed ten year strategy for transport in England. It set out a £180 billion investment package aimed at cutting congestion, providing more transport choice and encouraging greater integration. Of the projected investment, 59% was to be spent on public transport, including improvements to the national rail network, part funding of up to 25 new light rail routes in major cities, guided bus schemes, park and ride, priority routes and funding to improve rural transport.[7]

Although widely welcomed at the time, the 10-Year Plan was not without its critics who argued that projected spending was too dependent on investment from the private sector and revenue generated from congestion charging and workplace car parking, all of which has been overestimated. Other critics suggested that a commitment to build fifty new bypasses was a dilution of ITWP principles, while others such as Goodwin[8] also argued that it would be virtually impossible for the Government to hit its targets on congestion without demand management measures to restrict the overall growth in traffic levels. Although widely praised for its ambition and long-term commitments, some of these criticisms of the 10-Year Plan, as will be discussed below, were to prove well founded.

3 From Integration to Political Pragmatism: the Evolution of Transport Policy

If the creation of DETR and the publication of the Integrated Transport White Paper were expected to pave the way for the prosecution of a more integrated and environmentally conscious transport policy at national and local level, the subsequent evolution of transport policy—and the Ministries that oversee it—has merely highlighted the inherent political constraints involved in pursuing a sustainable programme. Within an unsympathetic political milieu, pragmatism appears to have increasingly prevailed over the approach set out in the ITWP.

There are a number of reasons why the Government has been unable to deliver fully on the White Paper, particularly in terms of environmental commitments. As Begg and Gray[3] suggest, institutional change, the emergence of congestion as

[7] Department for Transport, *A New Deal for Transport: Better for Everyone*, The Government's White Paper on the Future of Transport, 1998, available: http://www.dft.gov.uk /itwp/paper/index.htm. (accessed 1st October 2003).

[8] D. Begg and D. Gray The case for congestion charging, *World Economics*, 2002, **3** No. 3 July-September 71–83.

a more salient concern than the environment, increasing public dissatisfaction with a perceived lack of progress on transport, and several political shocks all combined to ease policy down a more pragmatic path.

In terms of institutional restructuring, despite its stated role of facilitating joined-up decision making, DETR was broken up in 2001. Transport and the Environment were split up; with the Department of Transport, Local Government and the Regions (DTLR) assuming responsibility for the former, while the Department of Environment, Food and Rural Affairs (DEFRA) took charge of environmental policy. In May 2002, DTLR was itself separated into the Department of Transport (DfT) and the Office of the Deputy Prime Minister (ODPM), following the resignation of Stephen Byers as Secretary of State for Transport. Within five years, the former DETR ministry was effectively divided into three departments, with ODPM in charge of housing, planning, urban policy and local and regional government.

It is logical to suggest that the fragmentation of the ministry responsible for integrated policy would make it more difficult for the Government to pursue a holistic approach. However, the institutional restructuring probably mirrored a shifting political perspective; a sustainable transport policy was—to a degree—expendable in the face of other, potentially more damaging, political pressures.

Labour inherited a situation whereby vehicle emissions were decreasing as a result of existing national and EU legislation, while car ownership, use and congestion were steadily increasing. Although tackling both vehicle emissions and congestion were twin pillars of the ITWP, progress on the former—as will be discussed below—was already something of a political success story.

On the contrary, traffic levels were still increasing, and the political reality was that congestion and the perceived inadequacies of alternatives to the car—such as rail—were emerging as more pressing concerns than vehicle emissions. Research carried out in 1998 by Sansom *et al.* demonstrated that the marginal cost of congestion per vehicle kilometre was around 10 p, around ten times that of local air pollution and approximately twenty times more than the cost of climate change.[9]

Given these trends—and the success of fiscal measures and voluntary agreements in cleaning up the UK's vehicles—it is not surprising that tackling congestion emerged as a priority as Government transport policy progressed through and beyond the *Transport Act* of 2000 and the subsequent 10-Year Plan for Transport.[10]

Increasing public dissatisfaction, a number of political shocks and the creation of a separate ministry for transport were also instrumental in the apparent shift away from the integrated approach set out in the ITWP. The UK's recent period of sustained economic growth followed at least two decades of under-investments in the country's transport infrastructure. In particular, the road network and the privatised and fragmented rail industry did not have the capacity to meet the increase in demand. As discussed below, the feeling that the Labour Government

[9] P. Goodwin, Infrastructure planning and forecasting: some current problems, *Conference on the Politics of Financing and Investment for Transport Infrastructure*, ECMT/TDIE, Paris, April 2003.

[10] T. Sansom, C. Nash, P. Mackie, J. Shires and P. Watkiss, *Surface Transport Costs and Charges*, 1998. DETR, London.

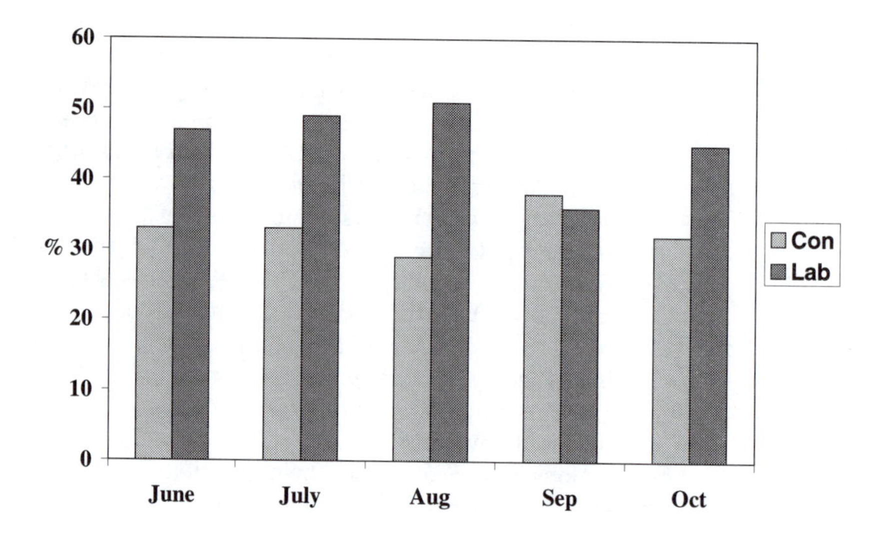

Figure 1 Government Popularity During the UK Fuel Protests. (Source: The Commission for Integrated Transport)

had failed to invest in transport (in transport infrastructure in particular) in its first term had the effect of creating widespread public dissatisfaction, which in turn led to a transport policy focused on infrastructure investment rather than the sustainable principles set out in the ITWP.

Public frustration over the Government's lack of progress on transport was brought into sharp focus by incidents such as the rail crashes at Ladbroke Grove and Hatfield, the collapse of Railtrack and the national fuel duty protest in September 2000. The impact of the fuel protests on transport policy was especially significant. As Figure 1 demonstrates, public dissatisfaction with the Government was such that, for the month of the protests, the Government fell behind the Conservatives in the opinion polls for the first time since they had come to power. In terms of voting trends, support for Labour fell from 51% to 36% while the Conservatives support increased by 9 percentage points to 38%.

This was a critical turning point for transport policy. Surprised by the strength of feeling over a transport issue, the Government realised that, not only had it to deliver (and be seen to deliver) on transport, it had to do so without further alienating a powerful motoring lobby.[3]

Significantly, the fallout from the fuel protests saw the end of the Fuel Duty Escalator, one of the Government's key sustainable transport instruments. The incoming Labour Government preserved the annual rise in fuel duty and then increased it to 6% in 1997. However, the cumulative effect of subsequent annual increases, magnified by rising world crude oil prices, resulted in the fuel price index increasing by 23% between January 1998 and July 2000.[11] Although the final price of fuel precipitated direct action, Glaister notes that the Government managed almost to halt traffic growth in the two years preceding the demonstrations and road use actually dipped during the peak in fuel prices and through the protests.[12]

[11] Department for Transport, *Transport 2010: The 10-Year Plan*, 2000, available http://www.dft.gov/uk/trans2010/plan/04.htm (accessed 1st October 2003).

Although the FDE was an effective demand management lever, it was fiscally crude and politically unpopular, and it was ultimately abandoned in the 2000 Budget. Duty was subsequently frozen in money terms (*i.e.* reduced in real terms) in the Budgets of 2001 and 2002 and increased in line with inflation in 2003. As a result — allowing for inflation — there has been a cut of around 6 pence per litre in petrol and diesel fuel duty rates since the fuel protest.[3] Glaister[12] estimates that, as a consequence, traffic will be about 1.8% higher than it would otherwise have been.

As noted above, with transport emerging as one of the key political issues of the day, delivering on transport — without further alienating the public — became a priority for the Government. Following the various bouts of departmental restructuring noted above, the Department of Transport was created in 2002, under Secretary of State for Transport, Alistair Darling. In contrast to DETR, where transport policy making was to be integrated with planning, regional policy and the environment, DfT was created as a singular ministry charged with '*bringing a clearer focus to transport thinking*'.[12] The implications for a sustainable transport policy are significant.

4 National Transport Policy

Since the publication of the ITWP in 1998, several variables have combined to shape the evolution of transport policy in the UK. As already noted, the 10-Year Plan and its key targets were criticised in some quarters as a dilution of the ITWP ideals. Recently, many of the goals on key indicators such as congestion have themselves been revised as the Government acknowledged that the 10-Year Plan was too optimistic.

In the *10-Year Plan Progress Report* published in December 2002, the original target (reducing congestion in England to below 2000 levels by 2010 across the inter-urban road network) was acknowledged as unachievable, vindicating earlier work by Goodwin. Instead, traffic is now forecast to rise between 20–25% over the ten year period, while congestion will rise between 11–20%.

This honest assessment of progress is typical of a more pragmatic regime that has characterized DfT following the appointment of Alistair Darling. The overriding need to deliver on transport and to avoid further controversy has seen Prescott's integrated approach set aside in favour of substantial investment in supply side infrastructure in the expectation that improvement will be delivered. Industry publications such as *Local Transport Today*[13] have also suggested that Darling was appointed with the aim of '*avoiding transport becoming a political hot potato again*'.

In analysing current transport policy in the UK, the Multi-Modal studies are a good example of how political pragmatism (combined with a lack of institutional integration) has undermined a policy instrument created to foster an integrated and sustainable transport system.

As a key element of Government transport planning, over twenty Multi-Modal studies have or are being carried out by consultants, who were charged with

[12] S. Glaister, UK Transport Policy 1997–2001, *Oxford Rev. Econ. Policy*, **18**, no. 2.

recommending the optimum transport solutions for a number of designated corridors prone to transport problems.[9] Most are inter-urban or peri-urban motorways. Nineteen studies had reported by March 2003. Although differing in approach, analysis and recommendations, a common theme among the various studies was an integrated 'package' of measures (including rail, road and demand management), mutually reliant elements that should not be introduced separately. Goodwin notes that several studies also concluded that managing demand for car use was essential, whether through road tolls or area wide congestion charging.

Recommendations for rail investment were subsequently greater than those for road building. Despite the integrated package proposed by many of the studies, Goodwin (Reference 9, p. 7) notes that '*the recommendations were unpacked, turned into list of projects, and then the projects referred to separate implementations agencies for further scrutiny, testing and decision.*'

In particular, the Strategic Rail Authority (SRA) — facing financial pressures and requiring its own strategic plan to improve the rail network — either rejected or were indifferent about the rail recommendation. In contrast, the Highways Agency was much more positive about the proposals for building or improving the roads. The Government for its part, were non-committal on the findings of those studies that emphasized the need for road tolls or charging.

As a result of the difficulty in signing up the SRA, and the Government's lack of commitment to the whole process, many of the programmes recommended in the studies will be implemented in isolation from other elements of the Multi-Modal 'package'. For instance, improvements in Multi-Modal corridors are included in a £21 billion package of trunk road improvements planned by DfT over ten years.[13] Although the projected increase in capacity is mirrored by a large increase in investment in public transport, the roads will undoubtedly generate additional traffic.

In July 2003, investment plans relating to eleven of the studies (covering the South Coast, West Midlands, Tyneside, South and West Yorkshire, Hull, the Thames Valley, the M25, the M60 around Manchester, and the corridors from London to Ipswich, Norwich to Peterborough and London to the South Midlands) were dominated by a £7 billion road building package.[14]

Environmentalists are concerned that road schemes are being unpacked from other measures for managing traffic levels recommended in the Multi-Modal Studies. Leaving aside the land required to widen and extend motorways, the two main issues are the impact on local air pollution and the production of CO_2.

In terms of the former, the widespread improvement in local air quality noted above has occurred almost independently of transport planning and traffic management policy. Emissions of pollutants that affect local air quality — such as nitrogen dioxide — have fallen dramatically as the average vehicle on the road becomes ever greener.

A series of voluntary agreements between vehicle manufactures and the EU, a cut in Vehicle Excise Duty to encourage the purchase of smaller cars and the inclusion of emission tests in MOTs have combined to clean up Britain's cars,

[13] Local Transport Today, Editorial, *Local Transport Today*, 2003, Issue 362, (20th March) 15.

[14] Department for Transport, *Delivering Better Transport: Progress Report*, 2003, available http://www.dft.gov.uk/trans2010/progress/ (accessed 1st October 2003).

while improvements to lorry fleets have been even more dramatic. Air quality emissions from each car are now one twentieth of what they were twenty years ago, while emissions of the most harmful local air pollutants from road transport (NO_x and PM_{10}) fell by 50% between 1990 and 2000, with a further reduction of around 30% forecast by 2010.[15] Buses are also much cleaner than they used to be. New buses and coaches must meet Euro III emission standards, and those placed in service since October 2001 are at least a third cleaner than those built under the previous Euro II standard introduced in 1993. Progressively tighter EURO IV and V standards will also come into force between 2006 and 2009.

As a result, targets on NO_x and PM_{10} for 2005 and 2010 are likely to be met over most of the country,[14] even against a background of rising car use. Although vehicle emissions have fallen, the growth in traffic means that there is more concern about future CO_2 emissions, which impacts on climate change. Since 1990, the average carbon efficiency of new cars entering the fleet has improved by around 10%, although the rise in car use over the same period means total carbon emissions have remained roughly the same.[16]

Measures in the 10-Year Plan were originally projected to save 3.8 mtC. This saving was expected to contribute to an overall reduction in greenhouse gases. However, the Sustainable Development Commission predict that, while the UK will be one of the few EU countries to meet its Kyoto target,[16,17] the domestic target—a saving 18.0 mtC of CO_2 emissions, or a 10.7% reduction from 1990 levels—is unlikely to be realised. The 10-Year Plan measures were expected to contribute 21% of the overall saving), but the Commission describes these savings as 'insecure'.

As noted above, the Government have also acknowledged this 'insecurity' in its 10-Year Plan Progress Report, recognizing that the traffic and congestion targets will not be met. In contrast, the House of Commons Select Committee Report on the Multi-Modal Studies reports that, at best, only 0.6 MtC will now be saved over the course of the 10-Year Plan and that carbon emissions may actually rise by 0.3 MtC (2003). Furthermore the Committee indicate that even allowing for technological improvements, carbon dioxide emissions are set to rise, possibly by as much as 2.05 MtC compared with 2000 levels. In addition, this does not include a reduction in traffic resulting from the introduction in area wide road-user charging, road tolls, or the other measures for restraining car use recommended by the Studies. Without these, the Select Committee suggests that carbon dioxide emissions could rise by 7.2 MtC.

It therefore appears that the current approach—in particular the absence of any mechanism to tackle increasing demand for car use—is unsustainable, especially with regard to future CO_2 targets.

[15] Department for Transport, *Announcement on Transport Investment by Rt Hon Alistair Darling*, 2003, available http://www.dft.gov.uk/stellent/groups/dft_about/documents/page/dft_about_022764.htm (accessed 24th July 2003).

[16] Department for Environment Food and Rural Affairs, *Our Energy Future—Creating a Low Carbon Economy Energy*, White Paper, 2003, available http://www.dti.gov.uk/energy/ white-paper/ourenergyfuture.pdf (accessed 1st October 2003).

[17] Sustainable Development Commission, *UK Climate Change Programme: a policy audit (SDC report)*, available http://www.sd-commission.gov.uk/pubs/ccp/sdc/index.htm (accessed 1st October 2003).

Although the Government has not explicitly rejected the integrated approach advocated in the ITWP, several influential commentators have noted the shift in emphasis and voiced their concerns. These include the ODPM Social Exclusion Unit,[18] the Commons Transport Select Committee,[19] ODPM Commons Select Committee[20] — who have noted a lack of integration between DfT and DEFRA — and the Commission for Integrated Transport.[21] In particular, CfIT has questioned how the Government can meet the targets set out in the ITWP if there is no management of demand for motoring. They also question how the behavioural change so strongly argued for in the 1998 White Paper is to be accomplished.

5 Local Transport Delivery

In delivering a sustainable transport policy, much currently depends on local authorities. In pursuing an infrastructure led approach, and attempting to avoid further alienating the powerful motoring lobby, the Government have been content to leave many of the difficult decisions on politically contentious issues — such as demand management and congestion charging — to councils and PTEs. In some respects this is far from ideal. Elected members and council officers face the same political pressures as national policy makers. Consequently, some are in a position to deliver sustainable policies, others are not.

A few local authorities have an exceptional record in pursuing a sustainable/integrated transport policy. For example, traffic levels in Nottingham and York have fallen against a background of continued economic growth,[22] through a combination of 'hard' and 'soft' measures that prioritise and promote other modes.

In an ideal world, all local authorities would be as forward thinking and tuned in to the integrated/sustainable transport agenda as York City and City of Nottingham Councils. Transport departments would be clearly delineated in function, efficiently run, well resourced and over-staffed with large numbers of eager, well-qualified officers. Meanwhile, their councillors would be well informed, far-sighted and courageous. Ruling groups would regard sustainable transport

[18] Sustainable Development Commission, *Policy audit of UK Climate Change Policies and Programmes (ECCM report)*, 2003, available http://www.sd-commission.gov.uk/pubs/ccp/eccm/index.htm (accessed 1st October 2003).

[19] ODPM Social Exclusion Unit, *Making the Connections: Final Report of Transport and Social Exclusion.* Social Exclusion Unit, Office of The Deputy Prime Minister, 2003, available http://www.socialexclusionunit.gov.uk/publications/reports/html/ transportfinal/office_d_pm.html (accessed 1st October 2003).

[20] House of Commons Transport Select Committee, *Transport: Third Report. Jam Tomorrow? The Multi Modal Study Investment Plans*, 2003, available http://www. publications.parliament.uk/pa/cm200203/cmselect/cmtran/38/3802.htm (accessed 1st October 2003).

[21] ODPM House of Commons Select Committee, House of Commons ODPM: Housing, Planning Local Government and the Regions Committee, *Department Annual Report and Estimates 2002.* Fifth Report of Session 2002-03, Vol. 1. available http://www.publications.parliament.uk/pa/cm200203/cmselect/cmodpm/78/78.pdf (accessed 1st October 2003).

[22] Commission for Integrated Transport, *10-Year Transport Plan Monitoring Strategy: Second Assessment Report*, 2003, available HTTP: http://www.cfit.gov.uk/reports/10year/second/index.htm (accessed 1st October 2003).

as a key priority. Their majorities would be large enough to allow them to push through potentially unpopular transport initiatives (such as demand management measures and congestion charging) in the face of local opposition, while still being confident of being around to enjoy the benefits in the long term. In such an ideal world, transport policy-making would be straightforward. Unfortunately, in the real world, only a small minority of authorities fit this profile.

In tackling congestion, for instance, while the Government has only recently started the debate on introducing measures to restrict traffic growth on the parts of the trunk road network that it controls,[23] over 90% of local authorities are critical of central Government for expecting them to introduce demand management measures in their areas.[24] However, opposition to measures to restrict car use in the media and among the public is such that even councils who see tackling congestion as a priority are likely to think twice before re-allocating road space away from the car or introducing charges. Many other authorities are even less likely to consider it. A survey of local authority transport officers found that only 26% believed that their Council had the will to tackle congestion.[25]

Councils are required to set out their transport strategy in Local Transport Plans (LTPs). LTPs are rated by DfT and used as a basis for allocating capital funding for transport schemes. However, the quality and content of LTPs varies considerably, and around a half of local authorities are either slightly behind or well behind in delivering their LTPs.[25] Local authorities are falling short of their spending because of a skills shortage, while revenue support for transport projects is also likely to be squeezed following the introduction of a single capital pot. Local Authority transport delivery is therefore likely to inhibit the Government's prosecution of a sustainable/integrated transport policy, particularly in regard to congestion.

Responsibility for local air pollution also rests with local authorities. Councils are required to designate Air Quality Management Areas (AQMAs) in localities where the Government's prescribed Air Quality Objectives are unlikely to be achieved; 114 AQMAs have been designated in the UK to date, the majority in respect of emissions from road transport (Reference 26, pp. 1–12).

Although some local authorities — such as Nottingham City Council — effectively integrate transport, land use and environmental policies to reduce local air pollution, evidence from elsewhere suggests that local integration is patchy at best.[19,23] In a survey of transport officers, around 37% thought that their authority had the political will to deliver on improving air quality.[25]

Despite the explicit link between air pollution and congestion, local authorities required to produce AQMAs are free to decide whether they integrate them with their LTPs.[27] Instead, and as noted above, the Government is relying on continued improvements to engine and fuel technology to hit its targets on local air quality.

[23] Commission for Integrated Transport, *Annual Report: 2001–2002*, 2003, available http://www.cfit.gov.uk/3annual/index.htm (accessed 2nd October 2003).

[24] Department for Transport, *Managing our Roads*, Department for Transport, London, 2003.

6 The Case for National Congestion Charging

Although the ITWP was widely welcomed by many academics and environmentalist, the pursuit of an integrated/sustainable transport agenda has been fraught. Although John Prescott was an enthusiastic advocate, DETR was perhaps too large and unwieldy to deliver, while Prescott was unable to sell the sustainable agenda to his cabinet colleagues, the key delivery agencies, and more than a handful of local authorities. Five years on, DETR has been broken up, and DfT appears to be pursuing a narrowly focused strategy.

Despite the successes of some local authorities, it can be argued that arresting the increase in traffic levels in general, and car use in particular, is ultimately the key to pursuing a sustainable transport policy in the UK (and elsewhere). Although more investment in public transport provision is necessary to provide high quality alternatives to the car, simply pouring money into bus and rail without managing demand for existing road space is not the answer. Even the highest projected increases in rail and bus transport will make only a few years difference to the growth of traffic.[5]

On the contrary, prioritizing a reduction in demand for car use will see other modes flourish, without the requirement for unsustainable levels of revenue support. However, since the abolition of the fuel duty escalator the Government has had no lever to pull to manage demand for car use. Without measures to manage traffic levels, the UK faces increasing congestion, higher levels of CO_2 emitted by vehicles, a potential increase in areas suffering from local air pollution, and other problems caused by rising traffic intensity such as social severance and noise.

Although the rise in car use needs to be arrested, it is unrealistic to expect local authorities to do this by themselves. Gray and Begg[6] note that, before they can introduce effective — but contentious — instruments such as congestion charging, local authorities must satisfy a number of economic, technical and political prerequisites. As a result, though congestion charging has been introduced successfully in London, similar large-scale schemes are unlikely to be introduced anywhere else in the UK, with the possible exception of Edinburgh. The 10-Year Plan forecast that twenty cities will have implemented some sort of charging system by 2010 (either congestion charging or work place parking), with charging introduced successfully in eight cities. Clearly this is not going to happen.

So what can be done? An unpublished report commissioned by DEFRA suggests that raising fuel duty by 10% and increasing Vehicle Excise Duty to £600 would be required to reduce CO_2 emissions from traffic.[25] Similarly, a report by the RAC Foundation suggests that for congestion to be held in check over next 30 years without congestion charging, the Government would have to build five times the amount of roads planned in the 10-Year Plan or increase fuel duty five-fold.

However, as stressed above, transport policy has to be delivered to a sceptical public, which consists of a motoring majority and a vocal environmental minority. Increasing the fiscal burden on the motorist by raising fuel duty or VED would

[25] Commission for Integrated Transport, *Local Authority Survey*, 2003 available HTTP:// www.cfit.gov.uk/ gov/reports/la/index.htm (accessed 1st October 2003).

be politically unpalatable, while a vast road building programme is neither feasible, desirable nor affordable.

Perhaps the only instrument that could both provide a more sustainable transport future, and which might be acceptable to politicians, the media and the public is national congestion charging, the merits of which are currently being debated in the UK. Proponents of national charging argue that, as well as easing congestion and cutting overall levels of traffic (and therefore vehicle emissions and the need for more road space), it is also equitable; motorists would pay for the congestion they cause. It might also be fiscally neutral (as will be discussed below), making it more acceptable, politically.

The current debate on national road user charging in the UK grew out a report published by the Commission for Integrated Transport in 2002. CfIT recommend scrapping Vehicle Excise Duty, reducing fuel duty, and charging motorists a variable rate for different types of road in different areas of the country at different times of the day. CfIT argue that this would bring a closer alignment between charges, environmental costs and other policy objectives on accessibility and social inclusion. For example, commuters travelling on congested roads at peak times would pay more than those using quiet roads at quieter times.

Not surprisingly, the optimal congestion charges estimated by CfIT would be highest in London, varying from 34 pence per km in Central London and 22 p/km^{-1} in inner London to 13 p/km^{-1} in outer London.[29] By contrast, there would be no charge on uncongested rural roads at off-peak times. For motorway travel, there would be an average charge of 2.2 per vehicle km for motorway travel, compared with 2.7 $p\ km^{-1}$ on other roads, although congestion charging would only apply on about 10% of the motorway network

CfIT estimate that national congestion charging would cut congestion in the UK by 44%, with traffic levels falling overall by 5%, and speeds increasing by 3%. The greatest benefits would also be experienced in the areas suffering the worst congestion. Incidence of gridlock would also be expected to fall significantly and reliability would improve. Journey times would get shorter and more predictable, which would benefit the economy as a whole and the road haulage industry in particular.

7 Delivering National Congestion Charging

In theory, a national congestion charging scheme appears to be an effective way of pursuing a sustainable transport policy. Although the Government are cautious about national charging, a recent seminar hosted by the Secretary of State for Transport suggests that the Government are at least willing to engage in the

26 Department for Environment Food and Rural Affairs, *Policy Guidance LAQM*, (PG3), 2003, available http://www.defra.gov.uk/environment/airquality/laqm/guidance/pdf/laqm-pg03.pdf (accessed 1st October 2003).

27 Department for Transport, *Guidance on LTP Annual Progress Reports* (3rd Edn.), 2003, Available http://www.localtransport.dft.gov.uk/Itp03/index.htm (accessed 1st October 2003).

28 S. Montague, Fuel price hike could cut pollution, *BBC News*, available http://news.bbc.co.uk/1/hi/uk/2989555.stm (accessed 1st May 2003).

29 Commission for Integrated Transport, *Paying for Road Use*, 2002, available HTTP: http://www.cfit.gov.uk/reports/pfru/index.htm (accessed 1st October 2003)

debate.[24] National road charging is endorsed by several national bodies, ranging from the RAC Foundation to the Commons Transport Select Committee.

Nevertheless, a workable national charging scheme is between six and ten years away and there are a number of significant considerations. Specifically, there are technical and political hurdles to overcome and doubts over whether the Government has the political will to impose a truly national scheme, or whether there are more acceptable compromises available, such as a voluntary scheme or a combination of urban road charging and motorway tolls.

Technical Considerations

Regarding the technology, there have been significant advances in the systems required to operate a national charging system in recent years (see Begg and Gray[8] and Clark[30]). It is likely that the system would be based on a Global Positioning System (GPS), linked to smartcard-charged units located in every vehicle. GPS could track each car to within 10 m across the entire road network, while an on board system would record information on trip duration, time and location.

Nevertheless, although much of the technology for such a scheme exists, its scale dwarfs the London congestion charging scheme in terms of complexity, the level of technology required and cost. It would be a substantial undertaking to introduce national charging, and it would have to work.

Another potential barrier is the conflict between privacy and enforcement. Clark[30] suggests that journey information could remain in-vehicle, guaranteeing anonymity—with the user paying by pre-pay smart card, direct debit or credit cards that are decremented to protect privacy. However, any system would have to be enforced, probably by a central computer, and potentially with additional camera enforcement. Ensuring that everyone pays, while accommodating the concerns of individual and human rights groups, will also be difficult, while the complexity involved in of kitting out 23 million cars with the appropriate technology should not be underestimated.

Political Considerations

In theory, a fiscally neutral charging system could deliver on environmental and social objectives without incurring the political backlash that is likely to accompany other fiscal measures or local congestion charging schemes outside London. In other words, while some motorists will pay more than they do now to drive on congested roads, others will pay less in charges than they currently contribute in fuel tax and VED. Overall, motorists would not contribute any more money to the Exchequer than they currently do. This is essential if the public, the media and politicians are to be convinced about the merits of a scheme.

However, in any shift in the burden of taxation there are likely to be new winners and losers. Although national charging would be more equitable, it is

[30] G. Clark, *Tax or Toll? Transport Research Foundation Fellowship Lecture*, Transport Research Foundation, Crowthorne, 2001.

likely to result in a significant regional shift in the fiscal burden. For instance, if 40% of congestion is in and around the M25, would those living in and around London and the south east be expected to pick up 40% of England's congestion charge? Would expenditure therefore have to be allocated in line with revenue generated to make a national charge politically deliverable? This would mean that motorists in rural regions would pay less in charges but would be—comparatively—starved of investment as a result.

Furthermore, motorists living in or travelling regularly to major conurbations may also pay more than they do now. As these are often traditional Labour heartlands, the Government would have to assess whether the benefits from a reduction in congestion outweigh the political ramifications of higher charges.

Consequently, it is likely that issues of 'equity' and 'fiscal neutrality' are a potential Pandora's Box, politically. Other considerations include the likely impact on traffic flows on quiet and rural roads that run close to trunk roads, and the extent to which cheaper motoring in rural areas will undermine rural bus services and hasten the decline in rural shops and services[8,30] also raises the question of whether such a scheme should be publicly and privately run and the obvious conflict between commercial objectives—maximizing revenue by increasing traffic—and Government objectives to limit or even reduce traffic growth. Furthermore, as has been demonstrated in London, if a national scheme proved too successful in discouraging car use, the potential loss of revenue might be a concern for the Treasury.

In weighing up all these factors, it is fair to conclude that, while a fiscally neutral national charging scheme currently commands widespread support, its delivery is by no means risk-free, politically.

Politically Acceptable Road Pricing: Alternative Scenarios. There are other models for road user charging that might be easier to implement than a national scheme. For instance, the Government could impose London-style congestion charging in other metropolitan areas and introduce tolling on motorways. This would have the advantage of being less expensive than full national charging, the technology is already proven, and it would avoid regional conflict over where revenues raised through national charging were spent. Whether a limited number of metropolitan schemes would reduce overall traffic levels across the network or simply move existing congestion elsewhere is less clear. However, partial or area charging is unlikely to be as effective in managing overall traffic levels and reducing CO_2 as a national scheme. Nevertheless, given the perceived success of charging in London, this could be the option favoured by the Government, especially once the potential difficulties emerge during the consultation process.

Another alternative is voluntary charging. Motorists who believe they currently pay too much could sign up to pay in the same way that households who consume the least water sign up to use water meters in England. It might be expected that rural and other low use motorists would sign up for national charging in return for not paying VED.

8 Conclusion

In providing funds for local authorities, investing in rail infrastructure and offering fiscal incentives for the uptake of cleaner vehicles, the UK Government has made more progress on sustainable transport than some critics have given it credit for. However, with recent policy focusing on increasing the supply of road space rather than constraining demand for the vehicles that will use it, the Government must now decide how it is going to manage traffic levels and congestion.

It is unfair to expect local authorities to make some of the politically difficult decisions on transport and to implement the effective — but vote losing — initiatives such as road charging. While some local authorities have been keen to implement a sustainable transport policy, they are very much in the minority.

In terms of effective instruments for delivering a sustainable transport, it has been argued that some form of nationally enforced road charging is required in the UK. The willingness of the Secretary of State for Transport to initiate a debate on the issue is welcomed.

If the intention *is* to pursue a sustainable transport policy, then national charging is a prize worth having. However, it would require a substantial amount of political will to deliver, and the many political, fiscal and technical hurdles standing in the way should not be underestimated.

As was suggested in the introduction, transport policymaking in the current social climate involves compromising philosophical ideals for the sake of political popularity. In the debate over national charging, it would be no surprise if a less contentious scheme (possibly voluntary or including metropolitan areas only) emerges as the favoured policy. The corollary is that the full environmental benefits of a national charging scheme are unlikely to be realized.

Water Pollution Impacts of Transport

DAVID MICHAEL REVITT

Introduction

This review concentrates on the sources, impacts and control options for aquatic pollutants derived from different forms of transport. The emphasis is on inland waters and hence there is no consideration of the possible polluting consequences of shipping on the marine environment. The effects of boating activities on inland waterways (*e.g.* rivers, canals and lakes) are mainly confined to recreational activities and are only referred to briefly. Contaminants that have been identified as being derived from boating traffic include aliphatic and aromatic hydrocarbons, polyaromatic hydrocarbons (PAH), methyl tertiary butyl ether (MTBE) (from fuels), copper (from anti-fouling paints), zinc (from sacrificial anodes) and fecal coliforms (from sewage discharges). An additional problem with boating activities is that in shallow waters those pollutants that tend to be accumulated in bottom sediments can be released into the overlying waters by mechanical disturbance of the sediment layer.

The main emphasis of this review is the widely studied impacts of road transport and its associated infrastructure on the quality of adjacent watercourses. The increasing concerns regarding the possible effects on receiving waters of the activities associated with the ground movements of aircraft are also fully discussed. The research literature associated with the impact of rail transport on water systems is less well developed and, therefore, is only mentioned where relevant pollutant usage has been shown to threaten water quality (*e.g.* the use of herbicides for track weed control).

Pollutants arising from all transport activities enter the water environment through normal usage practices but there can also be extreme pollution events such as those associated with spillages of either fuel or transported goods. Approximately 80 million tonnes of 'dangerous' goods are carried annually by road transport in the UK. There are relatively few major or serious incidents (between 50 to 100 per annum); most incidents being minor petrol spillages. It is not the intention to concentrate here on these infrequent pollution events,

Issues in Environmental Science and Technology, No. 20
Transport and the Environment

although the consequences may result in acute impacts and hence the need to have in place emergency control procedures needs to be recognized.

Many transport related activities (particularly those involving vehicles and aircraft and to a lesser extent trains) require the construction of extensive impervious areas which can result in modifications to the hydrological cycle. This may involve increased volumes of rainfall induced runoff being conveyed over shorter periods of time at elevated flow rates to adjacent surface waters and, in some cases, groundwaters. The impact of these episodic discharges can be acute in terms of both hydraulic and water quality terms. Engineering designs have consistently favoured the rapid removal of surface runoff from impermeable areas with the water quality effects being given secondary consideration. This approach is now changing with the recognition that surface discharges can pose serious detrimental impacts to receiving waters and, therefore, that control of runoff quality as well as volume is important. The water quality aspects will be concentrated on but reference to quantity issues will be necessary to recognize the role of pollutant loadings in contributing to the degradation of receiving water quality.

This review has three main sections, dealing with the sources, impacts, and treatment options for controlling transport related pollutants before they enter the water environment. The major sources of transport related pollutants are identified in Section 2 together with the environmental processes that can control their movement prior to entering the water system. In Section 3, the characteristics of the different groups of transport-derived pollutants are discussed together with the mechanisms by which they can exert an influence on receiving waters. Guidance on the best practical techniques currently available for controlling the discharge of transport related pollutants is provided in Section 4. The use of sustainable systems is becoming increasingly attractive, particularly where these enhance the ecological and aesthetic qualities of the local environment.

2 Sources of Transport-derived Pollutants

The main sources of pollutants arising from transport related activities are shown schematically in Figure 1 together with the processes by which they can ultimately be carried to the receiving water environment. The importance of the build-up of pollutants on impervious surfaces is clearly indicated as a regulating mechanism. Subsequently, during rainfall events, wash-off can occur to either the stormwater system prior to direct discharge to a receiving water or to the combined sewer system and eventually to the sewage treatment works. In some instances, infiltration leading to replenishment of groundwaters may be practised. Alternative removal processes include re-suspension under both dry and wet conditions as well anthropogenically controlled cleaning actions.

Pollutant accumulation rates have been widely studied on highway surfaces using particulate material as a representative parameter and have been based either on an exponential build-up during the dry weather period or on pollutant accumulation related to average daily traffic densities.[1,2] There is evidence that

[1] J. B. Ellis and D. M. Revitt, *Traffic Related Pollution on Highway Surfaces*, Urban

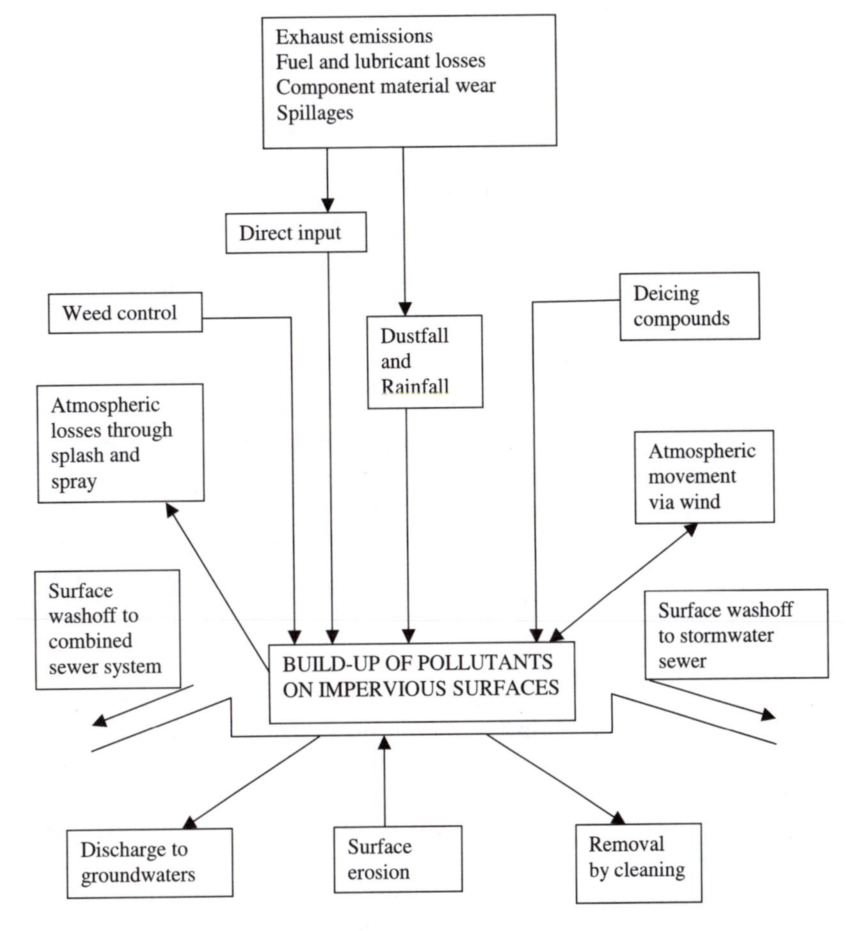

Figure 1 Pathways of transport-derived pollutants to the aquatic environment

for busy roads the surface material load eventually tends to stabilize around a constant level due to a balance between build-up and re-suspension due to turbulent eddies caused by fast moving traffic. A survey of European data on the accumulation of particulate material on highway surfaces indicates loading rates varying from a minimum of approximately 2000 kg ha^{-1} $year^{-1}$ for a residential road [average daily traffic density (ADT of <5000)] to over 10 000 kg ha^{-1} $year^{-1}$ for a urban motorway (ADT of $>50{,}000$). The wash-off of accumulated pollutants has commonly been modelled by an exponential relationship in which the rate of removal is assumed to be proportional to the amount remaining.[2] The importance of build-up and wash-off processes has been illustrated by an analysis of highway runoff data from Texas, USA in which the total suspended solids load was found to depend not only on the length of the antecedent dry period but also on the characteristics of the current and the preceding storms.[3]

Pollution Research Centre Research Report No. 18, Middlesex University, 1989.

2 V. Novotny and H. Olem, *Water Quality: Prevention, Identification and Management of Diffuse Pollution*, Van Nostrand Rheinhold, New York, 1994.

3 L. B. Irish, M. E. Barrett, J. F. Malina and R. J. Charbeneau, *J. Environ. Eng.—ASCE*,

Highway Environments

Pollutant inputs into the highway environment can be through direct inputs arising from vehicle emissions, vehicle part wear and vehicle leakages as well as by the carrying out of seasonal maintenance activities such as summer weed control and winter de-icing practices. In addition, the import of pollutants can occur through dry and wet atmospheric deposition processes (Figure 1). Up to 20% of total suspended solids and 10–50% of other constituents in highway runoff have been estimated to originate from atmospheric deposition during both dry and wet conditions.[4]

Vehicle Emissions, Vehicle Part Wear and Vehicle Leakages. Historically, the main concern relating to deposited material from petrol engined vehicle exhaust emissions has been the presence of fine particulate associated Pb compounds as a consequence of the use of tetra-alkyl Pb compounds as anti-knock agents.[5] The progressive phasing out of leaded petrol in the UK since 1987 and its banning from 1 January 2000 in most EU countries has significantly reduced the deposition of Pb on to highway and neighbouring surfaces and hence depleted the reservoir of this toxic metal available for wash-off into receiving waters. However, alternative additives, such as MTBE are now being used and there are concerns about the possible accumulation of MTBE in groundwaters.[6] Other organic pollutants present in vehicle exhaust emissions include uncombusted aliphatic hydrocarbons and polyaromatic hydrocarbons (PAH), which are produced as a result of chemical reactions within the engine and exhaust system. Leaks from lubrication and hydraulic systems to highway surfaces occur during normal vehicle operation, and Wixtrom and Brown[7] have reported running losses of total hydrocarbons in the range of 0.2 to 2.8 g per mile driven from this source. Used engine oil also contains metals such as Ba, Ca, Mg, Zn, Cu, Fe, Cd and Pb[8] with zinc dithiophosphate being added to motor oil as a stabilizing additive.

The main pollutants arising from the abrasion and corrosion of vehicle components are metals[9] although fine polymeric hydrocarbon particles (>0.01 μm) result from tyre wear. Lead oxides as well as Zn and Cd compounds are used as fillers in tyres. The replacement of asbestos in brake linings by high concentrations of Cu (*e.g.* 30 600 $\mu g\ g^{-1}$) provides added mechanical strength and assists heat dissipation. Other vehicle parts that produce metals due to erosion/abrasion processes include metal plating and bodywork (Cu, Cr, Fe and Ni), moving engine parts (Cu, Cr, Fe and Mn), and bearings/bushes (Pb, Cu and Ni).

1998, **124**(10), 987.

4 J. S. Wu, C. J. Allan, W. L. Saunders and J. B. Evett, *J. Environ. Eng.—ASCE*, 1998, **124**(7), 584.

5 P. D. E. Biggins and R. M. Harrison, *Environ. Sci. Technol.*, 1980, **14**, 336.

6 J. E. Reuter, B. C. Allen, R. C. Richards, J. F. Pankow, C. R. Goldman, R. L. Scholl and J. S. Seyfried, *Environ. Sci. Technol.*, 1998, **32**(23), 3666.

7 R. N. Wixtrom and S. L. Brown, *J. Exposure Anal. and Environ. Epidemiol.*, 1992, **2**(1), 23.

8 J. ZiebaPaulus, *Forensic Sci. Int.*, 1998, **91**(3), 171.

9 A. P. Davis, M. Shokouhian and S. B. Ni, *Chemosphere*, 2001, **44**(5), 997.

Road Surface Erosion. Three factors identified as influencing the generation of fine and coarse sedimentary materials from road surfaces are the age and condition of the surface, local climatic conditions, and fuel/oil leakages.[10] In northern European countries, the use of studded tyres in winter can cause extensive wear of road surfaces.[11] In addition to producing sediments, road surface wear can lead to the release of aromatic hydrocarbons and certain metals (particularly Ni) that are associated with the construction materials. Road markings contain metals such as Cr, Cu, Ti and Zn but unless the paint is in poor condition only a small contribution to road dusts should be expected.

Seasonal Maintenance Practices. *Winter maintenance practices.* In cold regions, snow handling measures such as ploughing and removal are essential to maintain safe road conditions. The environmental impact of these measures depends on the snow handling strategy employed and whether the snow, and its associated pollutants, are transported to a local or central deposit and dumped either on land or in a water body. Accumulated snow at the sides of busy roads is an efficient collector of inorganic pollutants derived from vehicular sources.[12] Within the snow deposits, pollutants are mainly particulate associated but transfer to the soluble phase can occur within the melt waters.[13] However, in densely developed urban centres, the soluble content of the melt is lower than in less densely developed residential areas due to preferential adsorption to the higher levels of particulates.[14,15]

The most commonly used de-icing agent on roads is a mixture of common salt and grit (loading rates of rock salt of 10 to 40 g m^{-2} are used in the UK depending on the severity of the freezing conditions). The presence of small amounts of hexacyanoferrate(II) ions to allow free-flowing applications of salt can result in the production of toxic cyanide ions due to photochemical decomposition.[16] Urea is sometimes used for bridge deck de-icing but presents the disadvantage that it is readily flushed into receiving waters, where hydrolysis to toxic, un-ionized ammonia occurs. A less corrosive alternative to sodium chloride is calcium magnesium acetate, which is less toxic to aquatic life and has a lower potential to participate in the mobilization of metals from roadside soils.[17]

Weed control. Most herbicides (typically 94%) used annually by local authorities in the UK are applied to roads and highways, parks, amenity grass and municipal paved areas. Within the highway environment, the purpose is to restrict unsightly

[10] J. D. Sartor and G. B. Boyd, *Water Pollution Aspects of Street Surface Contaminants*, US Environmental Protection Agency Report No. R2/72-081, Washington, 1972.

[11] A. Lindgren, *Sci. Total Environ.*, 1996, **190**, 281.

[12] E. L. Viskari, R. Rekila, S. Roy, O. Lehto, J. Ruuskanen and L. Karenlampi, *Environ. Pollut.*, 1997, **97**(1–2), 153.

[13] M. Viklander, *Sci. Total Environ.*, 1996, **189/190**, 379.

[14] J. J. Sansalone and S. G. Buchberger, *Trans. Res. Record*, 1996, **1523**, 147.

[15] M. Viklander, *Water Sci. Technol.*, 1999, **39**(12), 27.

[16] V. Novotny, D. Muehring, D. H. Zitomer, D. W. Smith and R. Facey, *Water Sci. Technol.*, 1998, **38**(10), 223.

[17] C. Amrhein, P. A. Mosher and J. E. Strong, *Soil Sci. Soc. Am. J.*, 1993, **57**(5), 1212.

weed growth and to reduce the possibility of structural damage occurring that may jeopardise safety requirements. Herbicides used include the triazines (atrazine and simazine), phenoxy acid compounds (2,4-D and mecoprop) and phenylurea compounds (diuron). Triazines were banned for non-agricultural use in 1993 and have been replaced by glyphosate and the additional use of diuron. The major factors that influence the removal of herbicides applied to hard surfaces by storm runoff are persistence, adsorption, rainfall intensity, and the time period between application and rainfall. In most circumstances, herbicide degradation will commence immediately due to the influences of photolysis and volatilization and, therefore, the ideal application programme will involve frequent dry weather applications of low doses.

Regular Maintenance Practices. Regular maintenance practices include road sweeping (mechanical brush sweeping and vacuum cleaning), the artificial flushing of pollutants from the road surface (hydrojetting), and the cleaning of gully pots. There are conflicting reports concerning the utility of street sweeping as a stormwater pollution control strategy although it clearly serves a 'cosmetic' function. The early work of Sartor *et al.*[18] found that conventional street cleaning practices were most effective in removing large particle sizes, with 70% removal of the sediment particles larger that 2000 μm compared with only 15% of those finer than 43 μm. Comprehensive studies in the USA under the National Urban Runoff Program[19] concluded that street sweeping could be an effective strategy for controlling stormwater pollution, particularly in climatic zones characterized by long dry periods. German[20] has commented on the elevated sediment yields associated with increased cleaning frequencies, although this was not in proportion to the increase in sweeping effort. Modelling studies have shown that the decrease in washed-off sediments depends exponentially on the sweeping frequency but that there is an insignificant increase in sweeping efficiency at frequencies of greater than twice per week.[21] The benefits of modern tandem street sweepers (brushing followed by vacuum) have been highlighted by Sutherland and Jelen,[22] with sweeping 26 times a year leading to a 18% reduction of suspended solids in wash-off. This technique can free surface-bound sediments during the brushing process, leading to higher sediment collection during the vacuuming phase.[23] The same authors have also studied the efficiency of hydrojetting on Paris streets, revealing highly variable efficiencies for solids

[18] J. D. Sartor, G. B. Boyd and F. J. Agardy, *J. Water Pollut. Control Fed.*, 1974, **46** (3), 458.

[19] J. D. Sartor and D. R. Gaboury, *Sci. Total Environ.*, 1984, **33**, 171.

[20] J. German, *Reducing Stormwater Pollution—Performance of Retention Ponds and Street Cleaning*, PhD Thesis, Chalmers University of Technology, Gothenburg, Sweden, 2003.

[21] A. Deletic, C. Maksimovic, F. Loughreit and D. Butler, *Proceedings of the 3rd International Conference on Innovative Technologies in Urban Storm Drainage*, Lyon, France, 1998, p. 415.

[22] R. C. Sutherland and S. L. Jelen, *Proceedings of the Stormwater and Water Quality Modelling Conference*, Toronto, Canada, 1995, p. 1.

[23] F-J. Bris, S. Garnaud, N. Apperry, A. Gonzalez, J-M. Mouchel, J-M. Chebbo and D. R. Thevenot, *Sci. Total Environ.*, 1999, **235**, 211.

(20–65%) and metals (0–75%) with no significant removal for PAH. This technique has the disadvantage that pollutants are transferred directly from the highway surface to the gully pot and the below-ground drainage system.

A comparison of the levels of pollution conveyed by rainfall and street sweeping operations *via* a bi-monthly vacuum cleaning programme in Bordeaux, France, indicated mass sediment removals of 55 and 45%, respectively.[24] However, because street sweeping is more effective in removing coarser material[18,25,26] the composition of the remaining sediment is changed towards a predominance of finer solids. It is with these fine particles that the higher pollutant concentrations are associated, and in the case of metals more than 50% are typically associated with fractions finer than 250 μm.[27–29] Investigations of the chemical associations of selected metals with the remaining street surface sediments suggest that the availability to the soluble phase of stormwater runoff is commonly in the order Cd > Zn > Pb > Cu.[30–32] The feasibility of re-using collected street sweeping waste has been questioned by studies in Sweden, where Zn and Cu levels in sediments accumulated over winter in the snow cover exceeded US EPA levels for dredged sediment disposal,[33] and in Scotland where road sweeping sludge has been characterized as a 'special waste' based on its oil and grease content.[34]

The total solids trapping efficiency of roadside gully pots (of which there are over 17 million in use within England and Wales) ranges between 15 and 95% depending on inflow, pot size and maintenance condition, but for particle sizes greater than 300 μm, the total solids reductions might be expected to be 70–75%.[35] During dry weather periods, rapid falls in dissolved oxygen can occur in the supernatant gully pot liquor, leading to anaerobic conditions and the release of soluble organics, ammoniacal compounds, dissolved metals[36] as well as sulfides. This toxic mixture is available for removal during the next storm event together with the accumulated sludge if the storm flow is sufficiently intense. Butler and Memon[37] have modelled the typical wet weather processes that occur within gully pots (including dilution, dispersion, sedimentation, wash-out of suspended and dissolved pollutants, and re-aeration). Clearly, the gully pot represents a

[24] P. Berga, *Optimisation du Nettoyage des Voiries*, Rapport CETE du Sud-Ouest, Ministere de l'Equipement, Bordeaux, 1998, p. 40.

[25] J. B. Ellis, in *Man's Impact on the Hydrological Cycle*, G. E. Hollis (ed.), GeoBooks, Norwich, UK, 1979, p. 199.

[26] J. German and G. Svensson, *Water Sci. Technol.*, 2002, **46**(6–7), 191.

[27] J. B. Ellis and D. M. Revitt, *Water, Air Soil Pollut.*, 1982, **17**, 87.

[28] J. J. Sansalone and T. Tribouillard, *Transport. Res. Record*, 1999, **1690**, 153.

[29] M. Stone and J. Marsalek, *Water, Air Soil Pollut.*, 1996, **87**, 149.

[30] R. M. Harrison, P. H. D. Laxen and S. J. Wilson, *Environ. Sci. Technol.*, 1981, **15**(11), 1378.

[31] R. S. Hamilton, D. M. Revitt and R. S. Warren, *Sci. Total Environ.*, 1984, **33**, 59.

[32] W. H. Wang, M. H. Wong, S. Leharne and B. Fisher, *Environ. Geochem. Health*, 1998, **20**(4), 185.

[33] M. Viklander, *J. Environ. Eng.*, 1998, **124**(8), 761.

[34] B. F. Clark, P. G. Smith, G. Neilson and R. M. Dinnie, *J. Chartered Instit. Water Environ. Manage.*, 2000, **14**(2), 99.

[35] D. Butler and S. H. P. G. Karunaratne, *Water Res.*, 1995, **29**(2), 719.

[36] G. M. Morrison, D. M. Revitt, J. B. Ellis, G. Svensson and P. Balmer, *Water Res.*, 1988, **22**(11), 1417.

significant source of pollution in the highway environment and can contribute to the 'first-flush' shock load of pollutants experienced by the receiving water. Regular gully pot cleaning is therefore needed to provide maximum protection to receiving waters and to reduce the probability of gully pots acting as sources rather than removers of pollutants in the highway environment.

Airport Environments

Glycols are extensively used within airports to prevent the icing of both runways and aircraft[38] but their cost inhibits widespread use in the highway environment. Ethylene and diethylene glycols are the principal components of de-icers that remove ice from impermeable surfaces during airport movements. Other formulations based on potassium acetate or urea have also been used as de-icers in the UK. Propylene glycol is widely used as the active constituent of anti-icers, which adhere to the treated aircraft surface and prevent the build-up of ice during taxi-ing and take-off. Polymeric thickening agents are added to retain the anti-icers on aircraft but it has been reported that up to 80% of the applied liquid runs off within the airfield boundary.[39] Another widely used group of additives are the benzotriazole derivatives as corrosion inhibitors, which increase the toxicity of de-icing fluids and decrease the potential biodegradation rate of propylene glycol.[40]

Railway Environments

There have been very few reports of the impact of rail transport on the aquatic environment. As for road transport, there is always a risk that the transport of hazardous and/or toxic chemicals can result in spillages that may reach adjacent water resources. Lacey and Cole[41] have commented on a protocol for estimating water pollution risks in relation to the flow of tanker wagons, the potential accident rate and the probability that an accident will result in a spill. Subsequently, it would be necessary to analyse the available surface water and also groundwater pathways to ascertain the possible impact to the aquatic environment and also contamination of drinking water sources.

Herbicides are applied by rail operators to the track areas to prevent excessive vegetation growth that could be detrimental to the safety of the railway system. Railway spraying operations have been shown to be the source of simazine in the runoff from an urban catchment bordered by a suburban railway line.[42] Leaching experiments on ballast removed from a railway line have shown that applied herbicides can persist longer than predicted by their respective soil-derived half-lives. This indicates the existence of a storage medium from which a gradual

[37] D. Butler and F. A. Memon, *Water Res.*, 1999, **33**(15), 3364.

[38] J. B. Ellis, D. M. Revitt and N. Llewellyn, *J. Chartered Instit. Water Environ. Manag.*, 1997, **11**, 170.

[39] R. O'Connor and K. Douglas, *New Sci.*, 1993, **137**(1856), 22.

[40] J. S. Cornell, D. A. Pillard and M. T. Hernandez, *Environ. Toxicol. Chem.*, 2000, **19**(16), 1465.

[41] R. F. Lacey and J. A. Cole, *Q. J. Eng. Geol. Hydrogeol.*, 2003, **36**(Part 2), 185.

release to both surface waters and groundwaters may occur under rainfall conditions a considerable time after application.[38]

3 Impacts of Transport-derived Pollutants

The ranges of pollutant concentrations measured in the drainage from highway and airport surfaces in the UK are listed in Table 1 together with event mean concentrations for specifically monitored storm events in the case of highway runoff. The quoted values indicate the wide variability that can exist, particularly for highway derived pollutants for which considerably more data are available. The data contained within Table 1 cover measurements taken during the last 30 years and, therefore, the quoted values may not exactly match those representing current transportation industry usage in terms of pollutant sources. This is particularly true for Pb concentrations as the values in Table 1 include those obtained prior to the reduction and subsequent banning of Pb in petrol. Motorways and trunk roads are defined in this instance by those highways carrying at least the equivalent of 30 000 vehicles per day. Within road environments, the pollutant loads flushed from the surface during rainfall events can be matched by the amounts dispersed by the spray and turbulence generated by passing traffic.[43] The latter effect will be greater on airport runways where the increased size and speed of aircraft can result in greater losses through this pathway. However, for roads, the distribution of pollutants due to spray and deflationary processes is limited to the immediate environment. Dierkes and Geiger[44] found the highest concentrations of heavy metals and hydrocarbons within the top 5 cm of roadside soils and within 2 m of the street. A subsequent rapid decrease with distance occurs, with backgound levels typically being achieved at 10–15 m.

Several researchers have investigated the impacts of the combined effects of the different pollutants found in highway runoff on receiving waters and come to different conclusions. In a study of nine East Anglian rivers, Perdikaki and Mason,[45] found that, in spite of seasonal variations, the ecological diversity did not differ significantly between sites upstream and downstream of road runoff discharges. In contrast, Marsalek *et al.*[46] compared the toxic effects produced by runoff from a highway carrying over 100 000 vehicles a day with those arising from urban stormwater using a battery of toxicity tests [*Daphnia magna*, Microtox (TM), sub-mitochondrial particles, and the SOS Chromotest] and observed that 20% of the highway samples were severely toxic compared with only 1% of the urban stormwater samples. Forrow and Maltby[47] used both *in situ* and laboratory studies to ascertain that reductions in the feeding rate of *Gammarus pulex* resulted from contact with contaminated sediments arising from toxic motorway

42 D. M. Revitt, J. B. Ellis and N. R. Llewellyn, *Urban Water*, 2001, **4**, 13.

43 J. B. Ellis and D. M. Revitt, *Drainage from Roads: Control and Treatment of Highway Runoff*, National Rivers Authority Report No. 43804/MID.012, Reading, Berkshire, UK, 1991.

44 C. Dierkes and W. F. Geiger, *Water Sci. Technol.*, 1999, **39**(2), 201.

45 K. Perdikaki and C. F. Mason, *Water Res.*, 1999, **33**(7), 1627.

46 J. Marsalek, Q. Rochfort, B. Brownlee, T. Mayer and M. Servos, *Water Sci. Technol.*, 1999, **39**(12), 33.

Table 1 Pollutant concentration ranges and event mean concentrations (EMC) in highway and airport drainage

	Highway runoff				Airport runoff
	Rural/suburban		Motorway/major highway		
Pollutant	EMC	Range	EMC	Range	Range
Suspended solids	41 mg l^{-1}	11–105 mg l^{-1}	261 mg l^{-1}	15–5700 mg l^{-1}	
BOD	17 mg l^{-1}	8–25 mg l^{-1}	24 mg l^{-1}	12–32 mg l^{-1}	10–4500 mg l^{-1}[a]
Chloride		1–27 mg l^{-1}	386 mg l^{-1}	80–2970 mg l^{-1}	
Total Zn	80 μg l^{-1}	20–1900 μg l^{-1}	41 μg l^{-1}	25–3550 μg l^{-1}	
Total Pb	70 μg l^{-1}	1–150 μg l^{-1}	96 μg l^{-1}	3–2410 μg l^{-1}	
Total Cu	40 μg l^{-1}	4–120 μg l^{-1}	150 μg l^{-1}	12–690 μg l^{-1}	
Oil/total hydrocarbons		3–31 mg l^{-1}	28 mg l^{-1}	8–400 mg l^{-1}	

[a] Refers to winter monitoring.

inputs. From a study of the characteristics of motorway runoff resulting from approximately 50 storm events, Legret and Pagotto[48] have defined two different types of pollution. The first type incorporating suspended solids, COD, total hydrocarbons, zinc and lead was referred to as chronic pollution, whereas seasonal impacts were identified with the second group, which consists of chlorides, sulfates, suspended solids and heavy metals resulting from the winter use of de-icing salts. Further details of the impacts of different types of pollutants found in both highway and airport runoff are described in the following sections.

Solids

The size ranges of suspended solids in highway runoff range from less than 1 μm to larger than 10 000 μm[49] with flow rate and storm duration being identified as the main controlling parameters with regard to the load and size of the transported solids. Laser particle sizing has shown that a significant proportion of the suspended solids load, on average 90% by weight, consists of particles finer than 100 μm.[50,51] On the road surface, solids predominantly collect adjacent to the kerb-side but they are also held within the pores of the road surface structure. This is where the finest fractions accumulate, and Furumai *et al.*[52] have postulated the occurrence of a stepwise wash-off phenomenon in which different wash-off behaviours are exhibited by the very fine (<20 μm) compared with coarser particles.

Highway-derived solids that are washed into receiving waters can exert detrimental ecological effects due to substrate smothering, reduction of light penetration due to increased turbidity, and a lowering of the oxygenation potential. These impacts are a function of particle size, which also influences, together with the mineral characteristics, the different affinities with which a variety of chemical pollutants, including metals and organic pollutants, can be attached to the surface of the particulate material. Roger *et al.*[51] have highlighted the relevance of motorway runoff derived suspended solids finer than 50 μm because of the increased surface area for pollutant adsorption and have indicated that excessive Pb and Zn contamination was observed in association with organic matter and clay contents. Following deposition within the receiving water environment, it is possible that weakly adsorbed toxic pollutants may be released into the aqueous phase and become more directly available for uptake by the existing flora and fauna.

[47] D. M. Forrow and L. Maltby, *Environ. Toxicol. Chem.*, 2000, **19**(8), 2100.
[48] M. Legret and C. Pagotto, *Sci. Total Environ.*, 1999, **235**(1–3), 143.
[49] J. J. Sansalone and S. G. Buchberger, *Water Sci. Technol.*, 1997, **36**(8–9), 155.
[50] D. Drapper, R. Tomlinson and P. Williams, *J. Environ. Eng.—ASCE*, 2000, **126**(4), 313.
[51] S. Roger, M. Montrejaud–Vignoles, M. C. Andral, L. Herremans and J. P. Fortune, *Water Res.*, 1998, **32**(4), 1119.
[52] H. Furumai, H. Balmer and M. Boller, *Water Sci. Technol.*, 2002, **46**(11–12), 413.

Metals

Numerous different metals have been reported to be present in road dusts, highway runoff, and in receiving water sediments immediately downstream of a highway discharge. Probably the most widely studied are Cd, Cu, Pb and Zn for which reported concentration ranges in road dusts are: non-detectable–11.4 $\mu g\ g^{-1}$, 25–1700, 35–10 700 and 96–3173 $\mu g\ g^{-1}$, respectively. In highway runoff a good correlation has been observed between suspended solids concentrations and particulate runoff concentrations for Cr, Cu and Zn,[53] and Krein and Schorer[54] have noted an inverse relationship between heavy metal concentrations and particle size in both road runoff and river bottom sediments. A similar observation was made by Sansalone and Buchberger[49] for Cu, Pb and Zn but not for Cd. Backstrom *et al.*[55] have compared the seasonal variations in concentrations of Cd, Co, Cu, Pb, W and Zn in the runoff from Swedish roads and found significant increases, of up to one order of magnitude, during the winter. This was particularly noticeable for Co and W and was considered to be due to a combination of the use of studded tyres and the chemical effects resulting from the use of de-icing salts. The platinum group elements (Pt, Pd and Rh) have been extensively studied in the past 10 years due to their increased use as catalysts in catalytic converters. Rauch *et al.*[56] have reported mean Pt, Pd and Rh concentrations in size fractionated road dusts ($<63\ \mu m$) of 341.3, 73.4 and 112.5 $ng\ g^{-1}$, respectively, for a busy urban highway. The event mean concentrations in highway runoff were calculated to be in the range 0.1 to 0.7 $ng\ l^{-1}$.[57] De Vos *et al.*[58] have extended the analysed range of PGE to include Ru, Os, and Ir and measured a maximum total PGE content in motorway runoff sediments of 55 $ng\ g^{-1}$.

The impact of metals in environmental samples depends on their partitioning between the particulate and dissolved phases, with the latter being more bioavailable and hence potentially more toxic. There are variable reports on the relative solubilities of different metals in highway runoff. Sansalone *et al.*[59] found that Cu, Cd, Zn, and Ni were mainly in the dissolved form, Cr and Pb were equally partitioned, whereas Al and Fe were strongly particulate bound. This latter trend was also observed by Shinya *et al.*[60] for urban highway runoff, although only Ni was found to be mainly in the dissolved form with the majority of the analysed metals (Cd, Cr, Cu, Mn, Pb and Zn) being mainly bound to particulates.

[53] M. Kayhanian, K. Murphy, L. Regenmorter and R. Haller, *Transport. Res. Record*, 2001, **1743**, 33.

[54] A. Krein and M. Schorer, *Water Res.*, 2000, **34**(16), 4110.

[55] M. Backstrom, U. Nilsson, K. Hakansson, B. Allard and S. Karlsson, *Water Air Soil Pollut.*, 2003, **147**(1–4), 343.

[56] S. Rauch, G. M. Morrison, M. Motelica–Heino, O. F. X. Donard and M. Muris, *Environ. Sci. Technol.*, 2000, **34**(15), 3119.

[57] C. Wei and G. M. Morrison, *Sci. Total Environ.*, 1994, **147**, 169.

[58] E. de Vos, S. J. Edwards, I. McDonald, D. S. Wray and P. J. Carey, *Appl. Geochem.*, 2002, **17**(8), 1115.

[59] J. J. Sansalone, S. G. Buchberger and S. R. AlAbed, *Sci. Total Environ.*, 1996, **190**, 371.

[60] M. Shinya, T. Tsuchinaga, M. Kitano, Y. Yamada and M. Ishikawa, *Water Sci. Technol.*, 2000, **42**(7–8), 201.

Equilibrium partitioning studies of the behaviour of Cd, Cu, Pb and Zn in stormwater suggest that the dominant water quality variables controlling soluble species formation are pH and alkalinity for Zn and Cd, with the less soluble Cu and Pb being strongly influenced by complexation with dissolved organic matter at pH levels below 8.[61] Revitt and Morrison[62] have determined the proportions of potentially bioavailable soluble metals in runoff from a car parking area for Cd, Cu, Pb, and Zn to be 59, 38, 5, and 53%, respectively, by applying a chemically defined speciation scheme to the soluble fraction.

The chemical speciation of particulate bound metals can strongly influence the ease with which they may be released into the soluble phase, thus providing information on the mobilization and transport of metals in the highway environment. The speciation scheme developed by Tessier *et al.*[63] has been widely applied to road dusts. This scheme involves five sequential chemical extractions, yielding fractions referred to as 'exchangeable', 'carbonate', 'reducible', 'organic' and 'crystalline lattice'. The 'exchangeable' fraction consists of metals loosely adsorbed through weak electrostatic attractions and which would be susceptible to desorption in the presence of high dissolved salt concentrations such as could exist following winter de-icing activities using rock salt. The 'carbonate' fraction contains metals associated with carbonate minerals, which can be released under increasingly acidic conditions. The 'reducible' fraction exists as nodules, concretions or cement between particles or as surface coatings due to metals adsorbed onto the oxides, hydroxides and hydrous oxides of Fe and Mn. These metals are prone to release when subjected to reducing conditions such as those that may occur upon contact with anoxic waters or waters with low redox potentials. Metals associated with the 'organic' and 'crystalline lattice' fractions are unlikely to be released under any changing conditions that are typically encountered in natural aquatic environments. Pb has been reported to demonstrate a high affinity (approximately 60% of the total content) for the 'exchangeable' fraction of road runoff particulates[62] and to some extent this balances the low 'bioavailable' soluble contribution found for this metal.

The ecotoxicological impact of a surface water outfall containing road runoff on two species of caged macroinvertebrates (*Gammarus pulex* and *Asellus aquaticus*) in the receiving waters has been studied by Mulliss *et al.*[64] Certain water quality parameters (BOD, suspended solids and total aqueous Cu concentrations) together with flow rate influenced the mortality responses of both species. Tissue concentrations of bioaccumulated Cd, Cu, Pb, and Zn were strongly correlated to the mortality response of *A. aquaticus* but this was not the case for caged gammarids for which survival was more dependent on total and dissolved metal concentrations. For platinum group elements, a bioavailability gradient of $Pt < Rh << Pd$ has been observed for *A. aquaticus* exposed to aqueous concentrations and to contaminated road sediments.[65] Exposure of the macroinvertebrates to 500 μg

[61] C. Dean, A. Blazier, E. Krielow, F. Cartledge, M. Tittlebaum and J. Sansalone, in *Global Solutions for Urban Drainage*, (*Best Management Practices*), E. W. Strecker and W. C. Huber (eds.), American Society of Civil Engineers, Reston, Virginia, 2002.

[62] D. M.Revitt and G. M. Morrison, *Environ. Technol. Lett.*, 1987, **8**, 373.

[63] A. Tessier, P. G. C. Campbell and M. Bisson, *Anal. Chem.*, 1979, **51**(7), 844.

[64] R. M. Mulliss, D. M. Revitt and R. B. E. Shutes, *Water Res.*, 1996, **30**(5), 1237.

l^{-1} standard solutions resulted in individual mortality rates of 47, 34 and 39% for Pd, Pt, and Rh, respectively. The importance of metal speciation with respect to bioaccumulation has been shown by comparing the behaviour of *A. aquaticus* with Pt(II) and Pt(IV) as chloride species. The preferential uptake of Pt(IV) suggested a charge-related bioaccumulation mechanism.[66] Laser ablation studies of the body of *A. aquaticus* exposed to road sediments contaminated with platinum group elements indicated that the metals were principally located at the head end, suggesting a rapid bioaccumulation mechanism.[67] This contrasts with the behaviour of other metals, such as Cu and Zn, which demonstrated a more uniform distribution along the body. Sures *et al.*[68] have advocated the use of the eoacanthocephalan parasite *Paratenuisentis ambigus* as a sentinel organism for very low environmental levels of Pt and Rh due to monitored bioconcentration factors of 50 and 1600, respectively, compared with water concentrations.

Inorganic Salts

Inorganic salts, such as nitrates and phosphates, may find their way onto highway surfaces due to the use of fertilisers in the adjacent environment. However, motorway runoff contains considerably lower phosphorus concentrations than general urban runoff, with the annual load of total phosphorus from a French motorway being estimated at 3.3 kg, of which 0.6 kg was in the form of orthophosphate[69]. Revitt *et al.*[70] found that orthophosphate was only infrequently detected in the routine monitoring of highway drainage over a 2-year period. In the same study, nitrate concentrations varied between 2 and 67 mg l^{-1} with the higher values observed immediately after the opening of a new road system when grassed verges had not become sufficiently established to efficiently utilize available nutrients.

Winter de-icing activities can introduce large amounts of chloride and, to a lesser extent, bromide and hexacyanoferrate(II) to the highway environment. The advantages of using salt have to be balanced against its environmental effects, which include the leaching of metals from basal sediments by ion-exchange processes and the depletion of dissolved oxygen levels due to enhanced stratification as a result of reduced vertical mixing. Because of these impacts, it has been recommended that the discharge of chloride to sensitive receiving waters should be carefully controlled.[71] The corrosive impact of chloride is indicated by

[65] M. Moldovan, S. Rauch, M. Gomez, M. A. Palacios and G. M. Morrison, *Water Res.*, 2001, **17**, 4175.

[66] S. Rauch and G. M. Morrison, *Sci. Total Environ.*, 1999, **235**(1–3), 261.

[67] M. Moldovan, S. Rauch, G. M. Morrison, M. M. Gomez, M. A. Palacios, ICP-MS and LA-ICP-MS as tools for the investigation of platinum group elements by the freshwater isopod *Asellus aquaticus*, Presentation at the *6th Winter Conference on Plasma Spectrochemistry*, Lillehammer, Norway, 2001.

[68] B. Sures, S. Zimmermann, C. Sonntag, D. Stuben and H. Traschewski, *Environ. Pollut.*, 2003, **122**(3), 401.

[69] M. Montrejaud Vignoles and L. Herremans, *Phosphorus Sulphur Silicon Relat. Elements*, 1996, **110**(1–4), 63.

[70] D. M. Revitt, R. B. E. Shutes, R. H. Jones, M. Forshaw and B. Winter, in *Proceedings of*

monitored vehicle corrosion rates that have been shown to be reduced by 50% on unsalted roads.[72] A possible consequence of continued high salt usage is an enhanced deposition of metals on to highway surfaces.

Despite the possibility of winter salt concentrations discharged from highway surfaces reaching high levels (Table 1) there have been no reports of increased chloride concentrations in British groundwaters.[73] This is not the case in Canada, where studies in the Greater Toronto Area, have shown that salting practices (at typical application rates of 200 g m^{-2}) have resulted in chloride levels of <2 to >1200 mg l^{-1} in groundwaters, of $>10\,000$ mg l^{-1} in shallow sub-surface waters and of >1000 mg l^{-1} in surface waters.[74] Earlier work in the same geographical area predicted an increase in spring water chloride levels to an unacceptable value of 426 ± 50 mg l^{-1} within a 20 year timeframe.[75] Chloride concentrations above 400 mg l^{-1} in receiving waters can stress sensitive fish and invertebrate species. However, the existence of high dilution ratios normally minimizes such impacts and prolonged chloride levels greater than 200 mg l^{-1}, the maximum acceptable value for drinking water, are unusual.

Organic Pollutants

Hydrocarbons are the largest group of organic pollutants found in the highway environment because of their association with the petrochemical products used in road construction (*e.g.* bitumen) and those derived from fuel combustion and engine additives. Kumata *et al.*[76] have recently identified tyre debris as a source of benzothiazolamines, which, because of their wash-off and sorptive behaviours, have been proposed as molecular markers for road runoff particles entering the aquatic environment. Because of their non-polar characteristics hydrocarbons become firmly attached to road sediments and remain in this condition when transferred to the aqueous environment with typically 70–75% of the total hydrocarbon load in highway discharges being associated with suspended solids. Total hydrocarbon levels in road runoff are shown in Table 1 with the elevated levels (up to 400 mg l^{-1}) having been recorded during short, intense storm events when suspended solids levels are high.[77]

The main types of hydrocarbons studied in highway discharges are aliphatic hydrocarbons (often referred to as oil and grease), aromatic hydrocarbons and PAH. The factors controlling the removal of the oil and grease component in highway runoff have been identified as runoff volume and number of vehicles during the storm event.[3] The strong association of PAH with suspended solids in

the 2nd National Conference on Sustainable Drainage, C. J. Pratt, J. W. Davies, A. P. Newman and J. L. Perry (eds.), Coventry University, UK, 2003, p. 19.

71 J. Marsalek, *Water Sci. Technol.*, 2003, **48**(9), 61.

72 B. Rendahl and S. Hedlund, *Mat. Performance*, 1991, **30**(5), 42.

73 M. Luker and K. Montague, *Control of Pollution from Highway Drainage Discharges*, Construction Industry Res. and Information Association, Report No. 142, London, 1994.

74 D. D. Williams, N. E. Williams and C. Yong, *Water Res.*, 2000, **34**(1), 127.

75 K. W. F. Howard and J. Haynes, *Geosci. Can.*, 1993, **20**(1), 1.

76 H. Kumata, J. Yamada, K. Masuda, H. Takada, Y. Sato, T. Sakurai and K. Fujiwara, *Environ. Sci. Technol.*, 2002, **36**(4), 702.

runoff has been clearly demonstrated and Krein and Schorer[54] have suggested a bimodal distribution in which the three-ring compounds are enriched in the fine sand fraction whereas the larger six-ring molecules predominate in the fine silt fraction. There is also an affinity of PAH for particulate organic carbon, particularly humic substances. When a 'first flush' phenomenon was observed for a highway runoff event, approximately 50% of the total PAH load was discharged and the predominant PAH in highway runoff have been consistently shown to be phenanthrene, fluoranthene and pyrene.[60] The total PAH concentration ranges for 16 monitored compounds achieved levels as high as 8150 ng l^{-1} and 1430 ng g^{-1} in the water and sediment of a road drainage ditch.[78]

Hydrocarbons can cause problems in receiving waters due to the build-up of a surface film, which can reduce the efficiency of oxygen transfer to the water body. Because of their affinity for the particulate phase, hydrocarbons tend to accumulate in bed sediments from which they are only slowly released even during times of sediment disturbance. Boxall and Maltby[79] have separated sediments contaminated with road runoff into three fractions of differing polarity and containing aliphatic hydrocarbons, 2–5-ring PAH, and substituted phenols together with 4- and 5-ring PAH. The middle fraction was found to be most toxic to *Gammarus pulex* whereas the first fraction was most toxic to *Photobacterium phosphoreum.* In a further refinement of this work, the same workers compared the toxicities of the three predominant PAH in highway runoff and showed that pyrene, fluoranthene and phenanthrene accounted for 44.9, 16 and 3.5% respectively, of the toxicity of a sediment extract.[80] Using the Ames assay, Shinya *et al.*[60] demonstrated that the mutagenicity of highway runoff was mainly associated with the particulate fraction although the dissolved fraction also showed a response that was identified with unknown soluble aromatic compounds.

The fuel additive MTBE is of significant concern because of its elevated solubility in water compared with other vehicle-derived organic compounds. Environmental levels of MTBE in the highway environment arise mainly from spillages and concentrations in the range 0.1–0.2 $\mu g\ l^{-1}$ have been recorded in groundwaters underlying motorways in SE England. Recreational boating is an important source of MTBE, giving rise to concentrations of between 0.1 and 12 $\mu g\ l^{-1}$ in a multiple-use lake.[6] The major loss of MTBE was by volatilization at the air–water interface, with the half-life varying from 193 days during the boating season to 14 days out of season.

The use of herbicides to control weed growth on highway surfaces and railway tracks has previously been described. These applications can eventually lead to the contamination of adjacent surface waters. A study by Revitt *et al.*[42] in an urban catchment containing both types of sources found elevated levels of several herbicides in the main drainage channel serving the catchment. The highest monitored concentrations were observed for diuron with a peak level of

[77] G. M. Colwill, C. J. Peters and R. Perry, *Water Quality of Motorway Runoff*, Transport and Road Research Laboratory, Supplementary Report No. 823, Crowthorne, Berkshire, UK, 1984.

[78] E. Naffrechoux, E. Combet, B. Fanget, L. Paturel and F. Berthier, *Polycycl. Aromatic Comp.*, 2000, **18**(2), 149.

[79] A. B. A. Boxall and L. Maltby, *Water Res.*, 1995, **29**(9), 2043.

248 $\mu g\ l^{-1}$ and an event mean concentration of 134 $\mu g\ l^{-1}$, corresponding to a 45% loss due to the impact of a spring rainfall event on diuron recently applied to the roadside verges. For simazine, the same storm event produced a peak concentration of 2.2 $\mu g\ l^{-1}$ and a percentage loss of 0.8% with respect to a railway application 262 days previously. This indicates the long-term persistence of simazine within railway ballast and its potential for slow release from this medium when exposed to intense rainfall events. Hence, despite the lower usage of herbicides in urban environments, these applications can frequently result in receiving water concentrations that are of regulatory concern, reaching concentrations well in excess of the 0.1 $\mu g\ l^{-1}$ drinking water limit for individual pesticides.

The high application rates of glycol based anti- and de-icers, which are used within airports during cold weather conditions, can pose serious problems for receiving water quality in the absence of treatment facilities for stormwater runoff. Glycols are completely miscible with water and can exert very high biochemical oxygen demands (Table 1), resulting in the degradation of water quality in receiving water bodies.[81] Glycols have been reported to have relatively low toxicities, with the green alga *Selenastrum capricornutum* demonstrating the most sensitive response when exposed to ethylene glycol (96 h LOEC of 1923 mg l^{-1}) in comparison to vertebrates and invertebrates.[82] However, aircraft de-icing fluids containing propylene glycol produced a 96 h LC50 of 18 mg l^{-1} for the fathead minnow *Pimephales promelas.* A separate study[83] showed that the LC50's of a similar anti-icer to a range of aquatic species were consistently less than 5000 mg l^{-1} propylene glycol although concentrations of up to 39 000 mg l^{-1} were observed at airport outfalls . Novak *et al.*[84] have tested the toxicity of a range of environmental samples (including stormwater) contaminated with aircraft de-icing fluids and found that 40% were lethal to rainbow trout and 30% were lethal to *Daphnia magna.* Potassium acetate is a less polluting, although more costly, alternative that exerts a smaller BOD than a glycol solution of equivalent de-icing power and has a lower toxicity. Urea is also non-toxic to fish but has the potential to exert a high biochemical oxygen demand and to be hydrolysed to ammonia[85–87] which in the un-ionized form is highly toxic to fish.

4 Control of Transport-derived Pollutants

The devices, practices or methods for removing, reducing, retarding or preventing

[80] A. B. A. Boxall and L. Maltby, *Arch. Environ. Contam. Toxicol.*, 1997, **33**(1), 9.
[81] G. S. Evans, *Proc. Institut. Civil Eng.—Transport*, 1996, **117**(3), 216.
[82] R. A. Kent, D. Anderson, P. Y. Caux and S. Teed, *Environ. Toxicol.*, 1999, **14**(5), 481.
[83] S. R. Corsi, D. W. Hall and S. W. Geiss, *Environ. Toxicol. Chem.*, 2001, **20**(7), 1483.
[84] L. J. Novak, K. Holtze, R. A. Kent, C. Jefferson and D. Anderson, *Environ. Toxicol. Chem.*, 2000, **19**(7), 1846.
[85] S. I. Hartwell, D. M. Jordahl and E. B. May, *Toxicity of Aircraft De-icer and Anti-icer Solutions on Aquatic Organisms*, Maryland Department of Natural Resources Final Report No. CBRM-TX-93-1, Annapolis, MD, 1993.
[86] D. A. Turnbull and J. R. Bevan, *Environ. Pollution*, 1995, **88**(3), 321.
[87] M. Koryak, L. J. Stafford, R. J. Reilly, R. H. Hoskin and M. H. Haberman, *J. Freshwater Ecol.*, 1998, **13**(3), 287.

pollutants in stormwater discharges from reaching receiving waters are generally described by the term 'Best Management Practices (BMPs)'. In the UK, the term 'Sustainable Drainage System (SUDS)' is commonly used to describe the sequence of management practices and control structures designed to treat impermeable surface runoff in a sustainable way. A difference between the two definitions is that BMPs also include non-structural aspects such as planning controls, educational aspects, regulatory controls, organizational/institutional controls and pollution prevention techniques. The latter includes street sweeping, which has been discussed above.

General design and technical guidance on BMP/SUDS have been provided in two Construction and Industry Research and Information Association (CIRIA) publications[88,89] together with a 'Best Practice Manual'.[90] There are also detailed design manuals on specific types of BMPs, including soakaways,[91] infiltration trenches,[92] constructed pervious surfaces[93] and constructed wetlands.[94] The Highways Agency have referred to the use of SUDS/BMPs for the control and treatment of highway runoff in the 'Water Quality and Drainage' section of the *Design Manual for Roads and Bridges*[95] and a more recent advice note specifically addresses the use of Vegetative Treatment Systems for Highway Runoff.[96] Applications of SUDS/BMPs are most extensively developed for highway drainage but there are some relevant examples for airport runoff. The following sections outline the principles behind the different structural BMPs, which can be used singly or in combination to attenuate the volumes, particularly the peak flows, of runoff derived from transport-related surfaces and to reduce the concentrations of pollutants associated with these discharges.

Filter Strips and Swales

Filter strips and swales are vegetated systems that can convey highway runoff

[88] CIRIA, *Sustainable Urban Drainage Systems: A Design Manual for Scotland and N. Ireland*, Construction Industry Research and Information Association, Report No. C521, London, 2000.

[89] CIRIA, *Sustainable Urban Drainage Systems: A Design Manual for England and Wales*, Construction Industry Research and Information Association, Report No. C522, London, 2000.

[90] CIRIA, *Sustainable Urban Drainage Systems: Best Practice Manual*, Construction Industry Research and Information Association, Report No. C523, London, 2001.

[91] BRE, *Soakaway Design*, Building Research Establishment, Digest No. 365, Watford, UK, 1991.

[92] CIRIA, *Infiltration Drainage: Manual of Good Practice*, Construction Industry Research and Information Association, Report No. 156, London, 1996.

[93] CIRIA, *Source Control using Constructed Pervious Surfaces*, Construction Industry Research and Information Association, Report No. C582, London, 2002.

[94] J. B. Ellis, R. B. E. Shutes and D. M. Revitt, *Constructed Wetlands and Links with Sustainable Drainage Systems*, Environment Agency, Report No. P2-159/TR1, Swindon, UK, 2003.

[95] Highways Agency, *Design Manual for Roads and Bridges*, HMSO, London, 1998, Vol. 11, Sect. 3, Part 10.

[96] Highways Agency, *Design Manual for Roads and Bridges*, HMSO, London, 2001, Vol. 4, Sect. 2, Part 1.

from the point of discharge and also provide storage and infiltration capabilities. In addition to attenuating peak flows and runoff volumes they are particularly effective at removing solids and associated pollutants through sedimentation, biofiltration and chemical adsorption. These processes are assisted by designs that ensure that sheet flow either across the filter strip or along the swale is achieved. These systems are particularly suited for use as source control treatments for runoff from small residential developments and parking areas. However, the use of swales as a 'first stage' treatment on a new dual carriageway road in Essex, UK, has recently been reported.[97]

Grass Swales. Critical parameters that influence the pollutant removal capacity of swales are the flow rates, length of swale and vegetation type and density. Vegetation resistance reduces flow velocities and increases contact opportunities between the flow and the vegetation and therefore enhances pollutant removal efficiency. Shallow and broad V-shaped grass channels are most efficient[98,99] with the range of pollutant removal efficiencies for highway runoff shown in Figure 2. Removal rates are less efficient for soluble metal species and nutrients[100] where longer residence times than the recommended minimum of 5 min would be beneficial. Concerns have been expressed about the build-up of pollutant deposits within the swales, but comparisons with pollutant loading criteria for the disposal of solids to land suggest that, with regular and appropriate maintenance, swales should have operational lifetimes in excess of 50 years.[101]

Filter Strips. Filter strips normally consist of sloping grassed areas and pollutant removal efficiencies are influenced by similar factors to those described for swales, with the longitudinal and cross slope gradients being particularly important. Barrett *et al.*[102] have recommended the use of filter strips with slopes of less than 12% and flow paths of at least 8 m for maximum pollutant removal effectiveness from highway runoff. The ranges of recorded removal efficiencies are shown in Figure 2, and the variability shown for filter strips is due to the high values reported for a 10 m grass filter strip receiving runoff from a heavy goods vehicle

[97] R. Macer-Wright, M. Brock and P. Ripton, in *Proceedings of the 2nd National Conference on Sustainable Drainage*, C. J. Pratt, J. W. Davies, A. P. Newman and J. L. Perry (eds.), Coventry University, UK, 2003, p. 71.

[98] P. M. Walsh, M. E. Barrett, J. F. Malina and R. J. Charbeneau, *Use of Vegetative Controls for Treatment of Highway Runoff*, Center for Research in water Resources, University of Texas, Report No. 97–5, Austin, Texas, 1997.

[99] G. W. Murfee, P. E. Scaief and J. M. Whelan, *Proceedings of 3rd International Conference on Diffuse Pollution*, Scottish Environmental Protection Agency, Edinburgh, 1999.

[100] C. Jeffries, G. M. McKissock, F. Logan, D. Gilmour and A. Aitken, *Preliminary Report on Swales in Scotland*, Scottish SUDS Working Party, Edinburgh, 1998.

[101] D. M. Revitt and J. B. Ellis in *Guidelines for the Environmental Management of Highways*, G. Mudge (ed.), The Institution of Highways and Transportation, 2001, p. 67.

[102] M. E. Barrett, P. M. Walsh, J. F. Malina and R. J. Charbeneau, *J. Environ. Eng.—ASCE*, 1998, **124**(11), 1121.

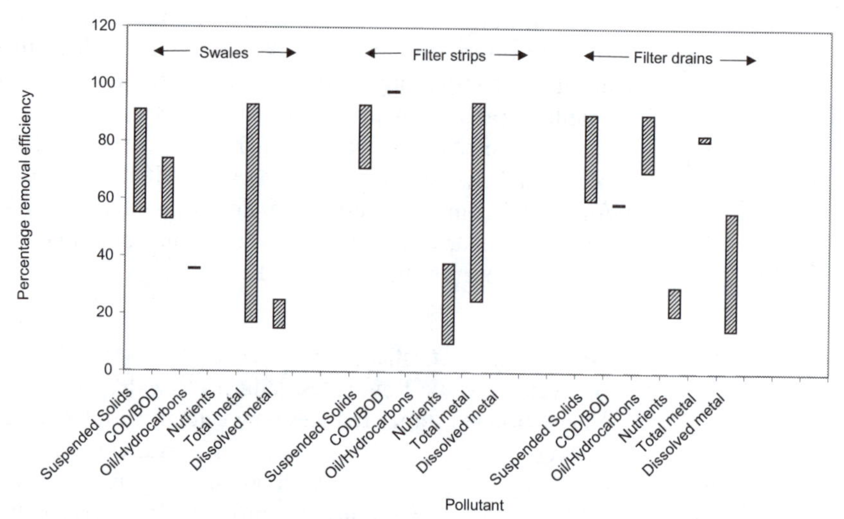

Figure 2 Ranges of pollutant removal efficiencies for swales, filter strips, and filter drains

car park located within a motorway service station near Oxford, UK.[103] Comparable studies in the USA have been less favourable[104] with the difficulties associated with nutrient removal being particularly obvious (Figure 2).

Filter Drains

Filter drains (also known as French drains) have been widely used in the UK to transport highway runoff to a suitable outlet point but there have been recent concerns regarding their tendency to undergo blocking by a combination of oil/grease and sedimentary material, leading to increased maintenance requirements and reducing operational lifetimes to approximately 10 years. Filter drains typically consist of perforated drainage pipes laid along the edge of highways in geotextile fabric lined trenches that are back-filled with granular material or lightweight aggregate. In addition to their runoff conveyancing function, they provide a treatment facility (through adsorption and biodegradation processes) and the average pollutant removal efficiencies observed for a 55 m length of filter drain adjacent to the M1 motorway near to Luton, UK[105] are shown in Figure 2. This diagram also shows the superimposed expected removal efficiencies for both conventional and toxic pollutants.[101] The variability in the potential for removing soluble metals is clearly indicated, with the highest value (56%) corresponding to zinc. There is also evidence of a limited removal potential for

[103] R. Bray, in *Proceedings of the 1st National Conference on Sustainable Drainage*, C. J. Pratt, J. W. Davies, A. P. Newman and J. L. Perry (eds.), Coventry University, UK, 2001, p. 58.

[104] S. L. Yu, W. K. Norris and D. C. Wyant, *Urban BMP Demonstration Project in the Albemarle/Charlottesville Area*, Virginia Department of Conservation and Historic Resources, University of Virginia, Final Report, Charlottesville, Virginia, 1987.

[105] R. Perry and A. E. McIntyre, in *Effects of Land Use on Freshwater*, J. F. Solbe (ed.), Ellis Horwood, Chichester, UK, 1986, p. 53.

[106] P. S. Mikkelsen, M. Hafliger, M. Ochs, P. Jacobsen, J. C. Tjell and M. Boller, *Water Sci. Technol.*, 1997, **36**(8–9), 325.

Table 2 Ranges of pollutant removal efficiencies for infiltration systems (soakaways, infiltration trenches, and infiltration basins)

Pollutant	*Range of percentage removal efficiencies*
Suspended solids	60–90
COD/BOD	70
Oil/hydrocarbons	70–90
Nutrients	20–50
Total metal	50–90
Dissolved metal	20–35

nutrients, represented by total nitrogen, which was also demonstrated by swales and filter strips.

Infiltration Systems

Examples of infiltration systems used for the treatment of transport derived pollutants include soakaways, infiltration trenches and infiltration basins. All these systems allow gradual infiltration of the contaminated runoff into the surrounding soils which must therefore possess pervious characteristics. Where there is the potential for polluting an underlying aquifer, a full risk assessment should be carried out to determine if interception is necessary. Improvements in water quality are achieved through physical filtration processes, adsorption of pollutants by the infiltration media (infill and surrounding soils) and microbiological removal of pollutants due to reactions on the surface of the media. The treatment efficiency depends on the contact time between the drainage waters and the infiltration media. Mikkelsen *et al.*[106] have demonstrated the effective pollutant trapping potential of these systems whilst also highlighting that a solid waste disposal problem could eventually result. The ranges of pollutant removal efficiencies that have been monitored for infiltration systems are shown in Table 2. The variability of the data is not unexpected given the size and scale of the infiltration systems for which results have been obtained. There is evidence that total metals are less efficiently removed than for previously discussed treatment systems but nutrients, as represented by total nitrogen, appear to have a slightly enhanced removal capability.

Soakaways. Soakaways are widely used in the UK to attenuate and treat highway runoff through gradual infiltration into the surrounding soil after passing through a coarse sediment trap and then either a chamber or stone-filled system. They have also been used for the treatment of airport runoff, with inputs of both glycols and urea in excess of 100 mg l^{-1} having been reported.[107] The same author has observed that pollutant levels peak at 0.4 to 0.5 m soil depths below the base of soakaways and subsequently decline exponentially to background levels. However, there are concerns about the behaviour of highly soluble contaminants such as certain Zn/Cd and cadmium species, selected herbicides

[107] M. Price, *J. Chartered Institut. Water Environ. Manag.*, 1994, **8**, 468.
[108] C. J. Pratt, *J. Chartered Institut. Water Environ. Manage.*, 1996, **10**, 47.

and MTBE, as tracer studies have indicated that trace levels could reach abstraction wells 3 km away from a soakaway injection point. The observed depth–concentration profiles suggest that pollutant decay rates are strongly influenced by both the available total organic carbon and the clay-silt percentages. Studies of the basal sediments in different soakaways have indicated the presence of high concentrations of total organic carbon and heavy metals associated with fine organic accumulations in the 400 mm directly below the soakaway.[108,109] The continued existence of this sludge layer exerts an important role in the retention of pollutants through filtration and sorption mechanisms.

Infiltration Trenches. Infiltration trenches operate in a similar manner to soakaways but require lower volumes of infiltration material (stone or rubble) for a given water inflow. Their lower popularity for the treatment of highway drainage, compared with soakaways, in the UK, relates to the commonly perceived design problem associated with accommodating the required trench length and width into the land area available. Studies of experimental systems in the USA have been carried out using different filler materials. A trench containing sand modified with an oxide coating was able to remove metals from highway runoff at efficiencies above 80% through adsorption–filtration mechanisms.[110]

Infiltration Basins. Infiltration basins are rarely used specifically for the treatment of highway runoff but are more likely to receive general urban runoff from mixed impervious surfaces. They can store runoff and subsequently allow it to percolate through either the soil base (sandy loams, sands, sandy gravels with an infiltration rate exceeding 15 mm hour^{-1}) or a specially constructed under-drainage system composed of gravel or sand filter beds. To achieve total solids removal efficiencies of up to 90%, an overall filtration rate of 5 m^{3} ha^{-1} m^{-2} is needed. Soils should also exhibit a high sorption capacity and a high resistance to desorption at low pH.[111] In common with other infiltration systems, a requirement in the design of infiltration basins is that one half of the total volume should be available within 24 hours of a runoff event and the maximum emptying time should be 96 hours.

Schueler[112] has studied a number of on-line and off-line infiltration basins in the USA and, while supporting the removal efficiencies quoted in Table 2, raises concerns regarding their long-term operational viability due observed failure rates of 50% within 5 years of installation. Factors such as lack of sediment pre-treatment, unsuitable soils, inadequate underdrainage and poor maintenance were cited as causes of failure and a cuurent view is that infiltration basins are best used as final polishing systems. Norrstrom and Jacks[113] found elevated

[109] J. Barker, W. Burgess, L. Fellman, T. Licha, J. McArthur and N. Robinson, *The Impact of Highway Drainage on Groundwater Quality*, Jackson Environ. Institute, University of East Anglia, Research Report No. 3, Norwich, 1999.

[110] J. J. Sansalone, J. A. Smithson and J. M. Koran, *Transport. Res. Record*, 1998, **1647**, 34.

[111] A. E. Barbosa and T. Hvitved-Jacobsen, *Sci. Total Environ.*, 1999, **235**(1–3), 151.

[112] T. Schueler, *Controlling Urban Runoff*, Metropolitan Washington Council of Governments, Washington, D.C., 1987.

[113] A. C. Norrstrom and G. Jacks, *Sci. Total Environ.*, 1998, **218**(2–3), 161.

concentrations of Cd, Pb, Cu, Zn and PAH in soils beneath an infiltration pond and incremental annual accumulations of Zn, Cu and Cd (averaging 0.8–1.5 mg kg^{-1} for Zn) were found in the basal sediments of an infiltration basin in Luton, UK, receiving peak discharges of 2.4 m^3 s^{-1} from a 26 ha residential site.[114] Exhaustion of the basin buffer capacity renders the long-term fixation of metals almost impossible, and, ideally, the top 30 cm of the infill material should be capable of maintaining pH in the range 5.5–8.0. This would reduce the probability of metal mobilisation and breakthrough due to the formation of soluble metal species. Where these are produced, the presence of a groundwater level more than 1 m below the base of the infiltration basin will allow the effective adsorption of colloidal metal species within the unsaturated zone.

Storage Facilities

These systems store the received surface runoff storm water prior to releasing it at an appropriate rate once the peak flow has passed. Water quality improvements are achieved by sedimentation, biodegradation and biological interactions (in vegetated systems). Examples of storage facilities used to treat highway runoff include sedimentation tanks and chambers and detention or retention ponds and basins as well as wetlands. The latter have also recently been incorporated into treatment systems for airport runoff in combination with balancing ponds containing assisted aeration.[115]

Storage Tanks/Chambers and Lagoons. Sedimentation tanks and chambers are artificial structures that may be built underground to reduce their visual impact. In contrast, lagoons consist of natural earth basins that may be covered with vegetation. Both treatment systems rely predominantly on the settling of solids for water quality improvements and the range of removal efficiencies observed in UK,[77,105] French[116] and German[117] studies are shown in Figure 3. Poor removal efficiencies for nutrients (represented by total nitrogen) and dissolved metals are exhibited once again. For most of the other studied pollutants the wide treatment ranges can be explained by the data for a lagoon system located adjacent to the M1 motorway, near Luton, which was considered to be over-designed.[105] Both types of treatment system require regular maintenance with major de-silting operations at intervals of approximately 5 years.

Detention and Retention Basins. Detention basins are dry, naturally vegetated impounding systems, essentially designed to provide flow attenuation. In extended

[114] J. B. Ellis, *J. Chartered Institut. Water and Environ. Manage.*, 2000, **14**(1), 27.

[115] D. M. Revitt, P. Worrall and D. Brewer, *Water Sci. Technol.*, 2001, **44**(11–12), 469.

[116] M. Ruperd, *Efficacite des Ouvrages de Traitement des Eaux de Ruisellement*, Service Tech. L'Urbanisme, Div Equip Urbains, Paris, 1987.

[117] G. Stotz, *Sci. Total Environ.*, 1990, **59**, 329.

[118] T. Petterson, J. German and G. Svensson, in *Proceedings of the 8th International Conference on Urban Storm Drainage*, I. B. Joliffe and J. E. Ball (eds.), The Institution of Engineers, Sydney, Australia, 1999, p. 1943.

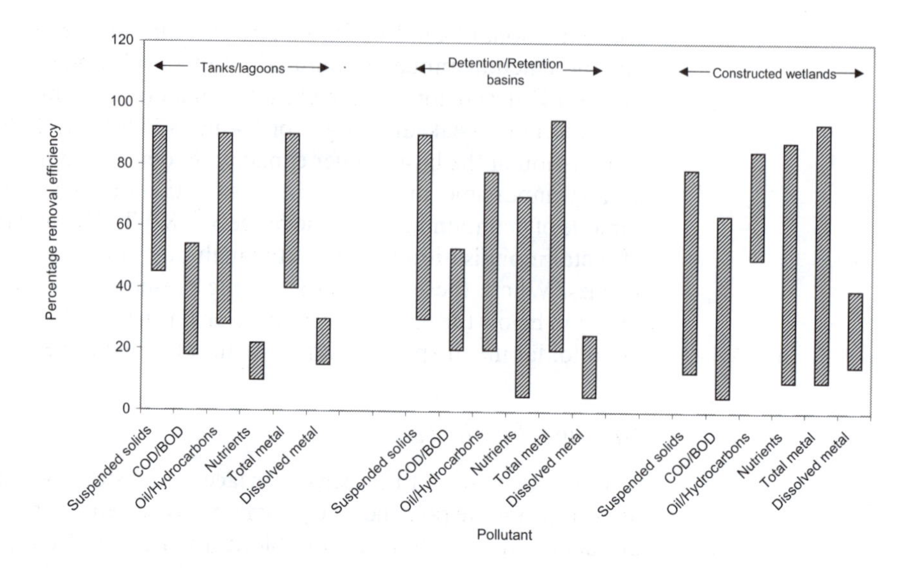

Figure 3 Ranges of pollutant removal efficiencies for storage facilities (tanks/lagoons, detention/retention basins, constructed wetlands)

detention ponds, the detention time is increased by up to 24 hours, providing an increased potential for the removal of fine suspended solids. Retention ponds/basins (or balancing ponds) are permanent water bodies that also provide increased runoff storage times and consequently offer increased treatment through the settlement of finer particles and also biodegradation of relevant pollutants.

The ranges of reported pollutant removal efficiencies for extended detention and retention basins are shown in Figure 3. The higher removal rates are generally associated with retention basins where a common practice of edge planting with emergent macrophytes provides additional biological treatment. This is particularly relevant for soluble pollutants and explains the enhanced removal efficiency for nutrients compared with other highway treatment systems. There is a distinct difference in the treatment capabilities for total and dissolved metals and this is exaggerated in cold climates, where metals in snowmelt runoff undergo preferential partitioning into the dissolved phase in the presence of elevated chloride concentrations.[118] In another study concentrating on metal removals, Hares and Ward[119] found elevated removal efficiencies (84–95%) for a 500 m^2 detention pond receiving runoff from the London Orbital M25 motorway following pre-treatment by a grit trap and an oil interceptor. However, it is essential to note that the variability in monitored removal efficiencies will also depend on specific design factors such as maximum water depth, aspect ratio (to prevent short-circuiting), storage volume compared with design storm volume, and the presence of inlet sediment traps and associated flow dissipation structures.[120]

Both detention and retention ponds require regular inspection and maintenance.

[119] R. J. Hares and N. I. Ward, *Sci. Total Environ.*, 1999, **235**(1–3), 169.

[120] M. J. Hall, D. L. Hockin and J. B. Ellis, *The Design of Flood Storage Reservoirs*, Butterworth-Heineman, London, 1994.

[121] Y. A. Yousef, D. M. Baker and T. Hvitved-Jacobsen, *Sci. Total Environ.*, 1996, **189**/**190**, 349.

The removal of contaminated sediments is required from detention ponds for both health and aesthetic reasons and from retention ponds to prevent damage to existing ecological systems. The build-up of metals in retention pond sediments has been studied by Yousef *et al.*,[121] and based on annual accumulation rates of 1.3, 13.8 and 6.9 kg ha^{-1} for Cu, Pb and Zn, respectively, removal of contaminated sediments every 25 years was recommended. Lee *et al.*,[122] found similar elevations of sedimentary heavy metal levels in French retention basins.

Constructed Wetlands. Constructed wetlands are fully vegetated systems that are normally designed to operate in either surface flow or sub-surface flow modes. They can provide the same level of treatment as retention ponds but over shorter periods of time and with reduced storage capacities. Sub-surface systems are becoming more popular and usually consist of gravel or artificial media containing emergent macrophytes such as *Typha latifolia* and *Phragmites australis*. These systems have been widely used for the treatment of sewage and for urban, industrial and agricultural runoff and are progressively finding applications for treating highway runoff[123] with currently a more limited usage for airport runoff.[115,124]

In sub-surface systems, the contaminated water flows horizontally or, exceptionally, vertically through the substrate and treatment occurs by several processes, including biofiltration, sedimentation, adsorption, biological uptake and physico-chemical interactions. Figure 3 indicates the range of pollutant removal efficiencies reported for constructed wetlands receiving highway runoff in the UK, France, Canada and the USA. The lower values in the quoted ranges are usually associated with the use of surface flow systems and, additionally, the variability in performance of both types has been attributed to short-circuiting, short detention and contact times, pollutant remobilization, and seasonal vegetation effects.[125] The performance of these systems with respect to nutrient removal is particularly dependent on seasonal factors. Constructed wetlands are generally efficient with regard to the reduction of the concentrations of particulate associated pollutants[126] and are efficient at trapping the fine (<63 μm) solids fraction.[127] The removal of soluble metals is less convincing and instances have been reported where higher metal concentrations exist at the outlet compared with the inlet due to flushing out of the wetland system during storm events.[70] This phenomenon was particularly evident for Cu, and does not appear to have a logical explanation as this metal would be expected to be strongly bound to sedimentary organic material.

122 P. K. Lee, J. C. Touray, P. Bailif and J. P. Ildefonse, *Sci. Total Environ.*, 1997, **201**(1), 1.
123 T. Bulc and A. Sajn Slak, *Water Sci. Technol.*, 2003, **48**(2), 315.
124 J. Higgins and M. Maclean, *Water Quality Res. J. Can.*, 2002, **37**(4), 785.
125 E. W. Strecker, E. D. Driscoll, P. E. Shelley and D. R. Gaboury, *Use of Wetlands for Stormwater Pollution Control*, US Environmental Protection Agency, 1992.
126 L. N. L. Scholes, R. B. E. Shutes, D. M. Revitt, D. Purchase and M. Forshaw, *Water Sci. Technol.*, 1999, **40**(3), 333.
127 H. Pontier, J. B. Williams and E. May, *Water Sci. Technol.*, 2001, **44**(11–12), 607.
128 J. Vymazal, *Water Sci. Technol.*, 1999, **40**(3), 133.

The application of constructed wetlands to the treatment of airport runoff has been mainly directed towards their ability to reduce the high potential BOD levels[128] associated with winter glycol applications. However, the use of pilot-scale reedbeds at London Heathrow Airport has demonstrated their potential for the removal of other airport-derived pollutants with maximum average removal efficiencies of 53, 64, 47, 35, 21, and 47% being monitored for nitrate,phosphate, ammonium, Cd, Cu, and Zn, respectively.[129] The average reductions in BOD concentrations were 31% for a surface flow system and 33% for a sub-surface flow system over a 2-year monitoring period. The corresponding average glycol removal efficiencies were 54 and 78%, following shock dosing inputs. Based on these results, a full-scale treatment system incorporating aeration ponds and a sub-surface flow constructed wetland has been built to receive the surface runoff from London Heathrow Airport,[130] and similar designs are being considered elsewhere.

There are no established design criteria for constructed wetlands for the treatment of highway runoff although this shortfall has recently been addressed.[94,131] Ideally, a wetland should retain the average annual storm volume for a minimum of 10–15 hours to achieve good pollutant removal efficiencies, but for good soluble metal removal efficiencies, and fine solids settlement a retention time of 24–36 hours would be preferred. However, in some instances for sub-surface systems, it may be practical to ensure that the first flush containing the heaviest pollution loads receives adequate treatment. Hydraulic retention time is a very important factor in the treatment performance of constructed wetlands with relevant influencing factors including the aspect ratio (width : length), the vegetation, substrate porosity and hence hydraulic conductivity, depth of water, and the slope of the bed. The recommended components of a constructed wetland treatment system for highway runoff would involve the following cellular structures arranged in series: oil separator and silt trap; spillage containment; settlement pond and associated control structures; constructed wetland and associated control structures; final settlement pond; and outfall into receiving watercourse. Regular sediment removal from the initial and final settlement ponds will be required and it is envisaged that the contaminated substrate within constructed wetlands will require cleaning or replacement to regenerate the hydraulic conductivity and pollutant removal capacity of the system after periods of approximately 25 years of operation.

Alternative Road Surfacings

Alternative structures based on the permeable paving principle allow water to pass through the surface into a pavement structure for temporary storage

129 D. M. Revitt, R. B. E. Shutes, N. R. Llewellyn and P. Worrall, *Water Sci. Technol.*, 1997, **44**(11–12), 469.

130 P. Worrall, D. M. Revitt, G. Prickett and D. Brewer, in *Wetlands and Remediation II*, K. W. Nehring and S. E. Brauning (eds.), Battelle Press, Ohio, 2002, p. 175.

131 J. B. Ellis, in *Impacts of Urban Growth on Surface and Groundwater Quality*, J. B. Ellis (eds.), Publication No. 259, IASH Press, Wallingford, UK, 1999, 357.

132 C. Stenmark, *Water Sci. Technol.*, 1995, **32** (1), 79.

followed by subsequent collection or disposal to the ground. The reductions in surface runoff are accompanied by improvements in water quality and hence less contamination of either surface or ground waters. Existing designs include porous asphalt, porous or solid blocks separated by voids (open or filled with permeable jointing material), and concrete grid pavers (containing large central voids filled with soil or gravel and often grass-seeded).

Porous Paving. Porous paving is ideally suited for use in driveways, residential cul-de-sacs, vehicle parking and service areas. It allows rainfall to infiltrate through pore spaces within the matrix of the material, as well as between individual blocks, into the sub-base. In this process surface runoff can be dramatically reduced and in some cases completely avoided, resulting in improved traffic safety due to enhanced skid resistance. Several researchers have reported that porous pavements are appropriate for the control of stormwater quantity and quality in cold climates, with the draining function preserved in snowmelt conditions.[132,133]

The percentage of water discharged from a 6250 m^2 car park area in a motorway service station was shown to be 33% lower than expected[134] and typical flow volume reductions were between 20 and 45% due to retention within the pavement reservoir. When these attenuations are combined with the pollutant reductions shown in Table 3 it is clear that there will be significant alleviations in the potential impacts on the receiving water environment. Pollutants are removed by a combination of filtration, adsorption and some sedimentation but Pratt *et al.*[135] have also shown that the sub-base of permeable pavements can act as effective *in situ* aerobic bioreactors, reducing oil-based contamination to less than 3% of that applied. Newman *et al.*[136] have shown that both oil and water retention can be improved by reinforcing the gravel packing in the reservoir structure with lightweight clay aggregates or porous concrete granules. Regular six-monthly cleaning, such as 'brush and suction' procedures, is required to prevent the clogging of surface porous materials. Where infiltration through surface voids is present, accumulations of silt and accompanying losses in permeability have been predicted to give a lifetime of around 15 years.

Porous Asphalt. Porous asphalt (or macadam) surfacing is composed of powdered/crushed stone with a bitumen binder to produce a coarse texture with a high void ratio. The open texture with continuous pore spaces allows immediate infiltration of rainfall. A hydraulic investigation of a 8000 m^2 supermarket car park covered with porous asphalt indicated an average reduction in outflow

133 M. Backstrom and M. Viklander, *J. Environ. Sci. Health*, 2000, **A35**(8), 1237.

134 C. L. Abbott, A. Weisgerber and B. Woods Ballard, in *Proceedings of the 2nd National Conference on Sustainable Drainage*, C. J. Pratt, J. W. Davies, A. P. Newman and J. L. Perry (eds.), Coventry University, UK, 2003, p. 101.

135 C. J. Pratt, A. P. Newman and P. C. Bond, *Water Sci. Technol.*, 1999, **39** (2), 103.

136 A. P. Newman, C. J. Pratt, S. J. Coupe and N. Cresswell, in *Proceedings of the 1st National Conference on Sustainable Drainage*, C. J. Pratt, J. W. Davies, A. P. Newman and J. L. Perry (eds.), Coventry University, UK, 2001, p. 425.

Table 3 Ranges of pollutant removal efficiencies for permeable paving systems (porous paving and porous asphalt)

Pollutant	*Range of percentage removal efficiencies*
Suspended solids	50–98
COD/BOD	36–89
Oil/hydrocarbons	70–98
Total metal	70–93

discharge of 51.5% over 20 monitored storm events.[134] On busy road systems there is an improvement in driver visibility through reduction of splash and spray and increased safety through a significantly reduced possibility of aquaplaning. The popularity of porous asphalt is also enhanced by the generation of less vehicle noise. Stotz and Krauth[137] demonstrated the ability of this type of surface to retain approximately 50% of the suspended solids in surface runoff. Legret and Colandini[138] have studied the behaviours of several different porous surfaces and observed percentage reductions in suspended solids of between 36 and 81%, with Pb removals of 76–86% and Zn removals of 35–67%. A particular problem with porous asphalt is the tendency of the pore spaces to become blocked by fine silt, reducing the efficiency of the infiltration process. This may occur within a period of 5 years, except where vehicle movements and pressures exerted by tyres counteract the build-up of silt.

5 Conclusions

Various pollutants, including solids, metals, inorganic salts, and fuel-derived organics are produced in the highway environment from direct vehicular sources and road surfaces as well as the routine and seasonal regimes associated with maintenance activities. Studies of airport environments have concentrated on the pollutants produced during winter de-icing activities and their potential oxygen demands on receiving waters. Within the railway environment, only herbicide pathways have been discussed in any detail in relation to the potential toxic impacts on receiving waters.

Combinations of *in situ* and laboratory studies have been conducted to assess the toxicity of transport-derived aquatic pollutants. Ecotoxicological studies have noted the relationships between mortality rates for macroinvertebrates and both metal body burdens and soluble metal concentrations. The distribution of metals in highway runoff depends on their affinity for particulates, with the major chemical influences being pH, alkanity and dissolved organic matter, and, in winter, the existence of elevated chloride levels. Organic pollutants identified in waters receiving highway runoff include hydrocarbons, MTBE and pesticides. Particulate associated PAH, with pyrene, fluoranthene and phenanthrene being dominant, are particularly toxic to native macroinvertebrate species. In the airport environment, the high BOD loadings generated by winter discharges of glycol based de-icers result in the degradation of water quality in receiving water

[137] G. Stotz and K. Krauth, *Sci. Total Environ.*, 1994, **146/147**, 465.
[138] M. Legret and V. Colandini, *Water Sci. Technol.*, 1999, **39**(2), 111.

bodies although there are varying reports on the direct toxicities of glycols.

A range of conventional systems and sustainable drainage systems (SUDS) are available to treat contaminated highway discharges. Different street-sweeping techniques are not efficient at removing the finest fractions of surface sediments with which the highest concentrations of metals and PAH are associated. Gully pots can trap a range of sediment sizes but, without regular cleaning, the supernatant liquor can undergo anaerobic reactions leading to the eventual release of toxic pollutants to the receiving waters. The range of SUDS applied to the control of highway runoff have been categorized as filter strips and swales, filter drains, infiltration systems, storage facilities, and alternative road surfacings. Variable pollutant removal efficiencies have been reported, with nutrients and dissolved metals being the most difficult to remove. Constructed wetlands demonstrate wide ranges and exhibit high pollutant removals in comparison with other treatment systems with maximum percentage efficiencies generally being in excess of 90% although for dissolved metals a maximum of only 40% is achieved. These systems have also been successfully used for glycol removal from airport runoff.

Climatic Impact of Surface Transport

MARTIN G. SCHULTZ, JOHANN FEICHTER
AND JACQUES LEONARDI

1 Historical Evolution of Surface Transport

Mobility is one of the most important factors for modern industrial societies in the globalized economy of today. About 90% of the total passenger transport (in passenger km) and 40% of the freight transport (in tonne-km) occurs at the surface, and most of this is powered by fossil fuel combustion engines. The average distance per person and the number of freight tonne-kilometres per year has increased steadily since the industrial production of the automobile in 1910 (Table 1 and Figure 1). Traffic is increasing at a faster rate than the global economy, population, or energy consumption.[1] During the 20th century, the world population grew by a factor of 4, while motorized traffic increased by more than a factor of 100. Despite economic recession in several areas of the world, mobility has continued to increase throughout the past decade.

There are various reasons for the tremendous increase in surface traffic: industrialization and motorization in developing countries, the global distribution of industrial production, technological innovation, and the continuous growth of worldwide trade all play a key role. Increased per capita income in several developing countries makes it possible for ever more people to afford a car or to travel by air plane, train, or ship. The increase in population furthermore enhances the demand for transportation. Last, but not least, the way our cities and settlements are built is increasingly oriented toward the use of cars and trucks as the main means of transport for passengers and freight.

Among the different traffic sectors, it is clearly road traffic that has driven this growth rate and which is responsible for the strong increase in fossil fuel consumption (Figure 2). The global automobile fleet was estimated to be about 508 million cars in 1998. In the US, two out of three persons own a car, in Europe this ratio varies from two to three persons per car. In China and India there were still less than 1 car per hundred inhabitants around 1995. Nevertheless, China

[1] R. Gilbert, *Sustainable Mobility in the City*, Centre for Sustainable Transportation, Toronto (www.cstctd.org/CSTadobefiles/sustainablemobility.pdf), 2001.

Issues in Environmental Science and Technology, No. 20
Transport and the Environment

Figure 1 Growth of global freight and passenger transport and global population increase (Reference 1, p. 12)

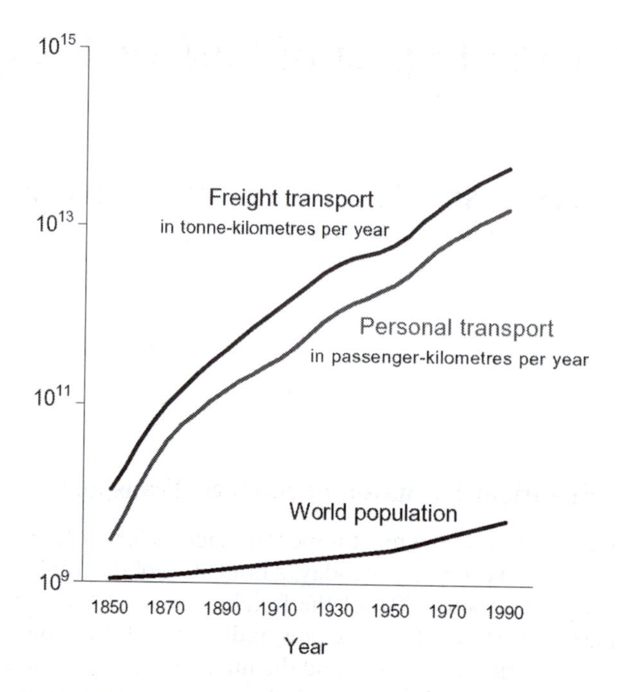

Table 1 Growth of global freight and passenger transport and global population increase (Reference 1, p. 12)

Year	*Freight transport (10^9 metric tonne-km)*	*Passenger transport (10^9 passenger-km)*	*World population (10^9)*
1850	10	5	1
1870	100		
1890	550	100	1.8
1910	1500		
1930	7500		
1950	8000		3
1970	12 000		
1990	60 000	12 000	5.6

has seen a tenfold increase in the number of automobiles between 1970 and 1990 (Table 2). Consistent with this increase, the average mileage per person each year has also increased over time (Figure 3).

In 1996, road traffic accounted for the consumption of about 1300 Mt oil equivalent, aircraft used about 200 Mt, and ship and rail traffic about 50 Mt each.[2] Ships clearly had the largest share of global freight traffic with about 3.8 $\times$ 10^{13} tonne-kilometres (tkm) in 1999, equivalent to 6300 tkm per person per year.[3] During the past decade, almost all major sea ports have seen a twofold increase in shipments, mainly from containers (Table 3). In 2002 alone, Shanghai experienced a 43% increase in container traffic volume. Container transport still

[2] WRI/UNEP/WBCSD (World Resource Institute/United Nations Environment Programme/World Business Council on Sustainable Development), *Tomorrow's Markets. Global Trends and their Implications for Business, Washington*, 2002.

[3] OECD EST — *Environmentally Sustainable Transport*, Organisation for Economic Co-operation and Development, Paris, 2001.

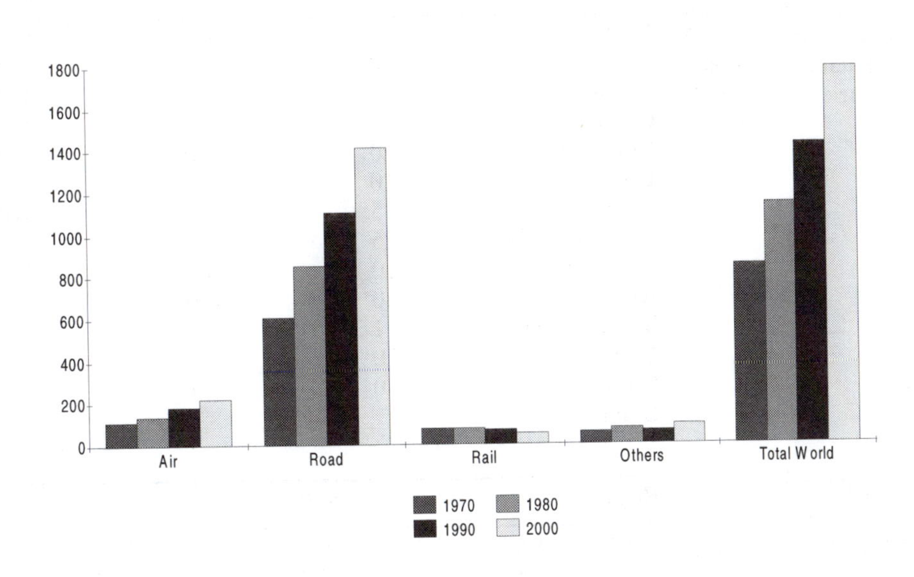

Figure 2 Modal split and total final world-wide energy consumption by the transport sector, 1970 to 2000, in million tons oil equivalent (Reference 4). 'Others' include estimates for pipelines, waterway transport and international marine bunker

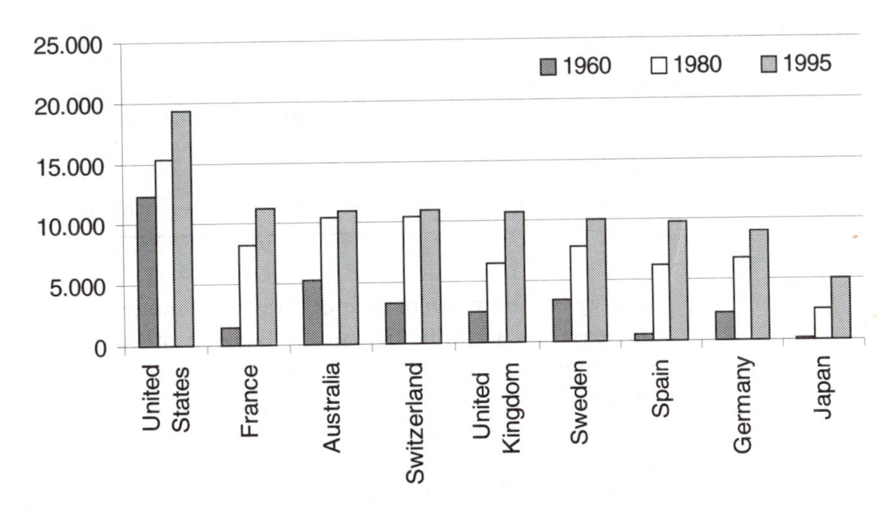

Figure 3 Change in the average automobile distance travelled per person each year; mean values for OECD countries (Reference 2, p. 3–3)

continues to be a major growing transport sector, with more than 10% increase per year in the last 5 years in all the major parts of the world.

Table 4 lists the consumption of different fuel types for traffic use in 1973 and 2001. While coal use has declined by almost three quarters, oil consumption has increased by a factor of 1.7, and electricity and natural gas have experienced growth rates of a factor of 2. However, in absolute numbers, oil consumption far outweighs other fuel types. In 1973, about 95% of all traffic was powered by oil, and by 2001 this fraction had increased to 96%. This growing dependency on oil is clearly due to the increase in road traffic and the parallel decline of rail transportation, both for passengers and freight. This is reflected in the development of the modal split (Figure 2). Comparing the magnitude and dynamics of the

Table 2 Historical development of the average number of vehicles per 1000 inhabitants for selected countries (Reference 2, p. 3–3)

Country	*1960*	*1980*	*1995*
USA	350	610	670
Italy	40	320	550
Germany	65	330	495
Canada	230	420	480
France	120	340	430
UK	110	280	380
Japan	5	180	310
Czech Republic	25	150	320
Hungary	10	80	260
Argentina	40	170	210
South Korea	1	15	180
Brasilia	20	70	100
China	0	1	8
India	1	2	7

Table 3 Change in container turnover in different regions of the world between 1980 and 1998 in thousand TEU (twenty feet equivalent unit) per year (Reference 2, p. 6–7)

Region	*1980*	*1998*
North America	9000	26.000
Western Europe	12000	43.000
East Asia	7500	50.000
South East Asia	2000	27.500
Middle East	2000	8.500
Central America	2500	14.000
Oceania	2000	4.500
South Asia	750	5.000
Africa	2000	5.750
Eastern Europe	500	1.500

energy consumption of different transport modes, it is clearly dominated by the increase in car traffic (Figure 2). Because of the direct causality between fuel consumption and CO_2 emissions, and the strong relationship between traffic performance and fuel consumption, it is worth looking closer at the historical development of traffic performance and estimating the future trends in the different traffic modes.

Reliable statistical data on the performance of different transport modes are, unfortunately, only available for OECD member countries and for countries from the CIS (Community of Independent States), but they confirm the above estimates of energy consumption.[4] Global estimates are incomplete and end in 1990.[1] Figure 4(a) shows that the total distance travelled on roads (cars, buses, and coaches) increased by almost a factor of two for passengers. Figure 4(b) shows the new OECD data on modal split performance for freight. Rail traffic has experienced a steady decline since 1970, and a rapid global downward trend since 1990, mainly because of the reduced freight transport volume in Russian railways. Nevertheless, about 17% of land-based passenger traffic and 40% of

[4] OECD *Environmental Data, Compendium, Transport*, Organisation for Economic Co-operation and Development, Paris, 2002.

Table 4 Contribution of different fuel types to global fuel consumption by traffic in 1973 and 2001 in Mt oil equivalent (ref. 15, p. 39)

Fuel type	*1973*	*2001*
Coal	26	5
Oil	903	1715
Natural gas	17	54
Others	10	27

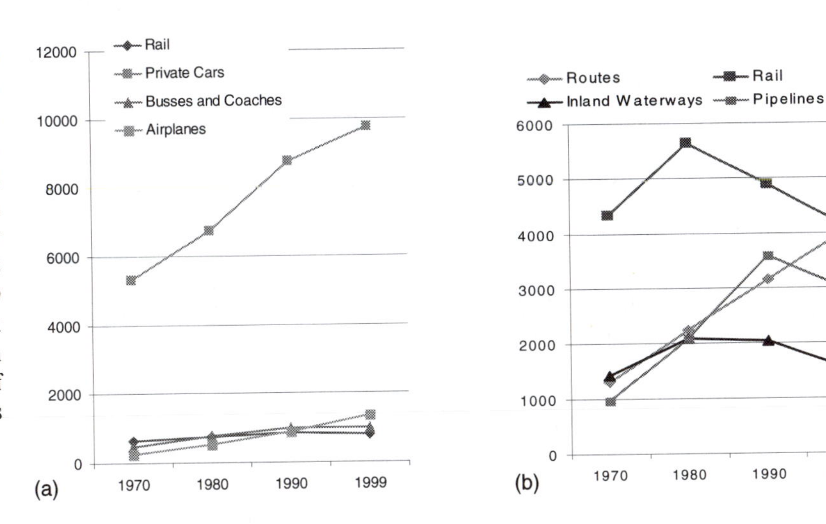

Figure 4 Historical development of the OECD and CIS countries total distance travelled in billion passenger km (a), and total freight transport in billion ton km (b) aggregated by transportation means (Reference 4). The OECD trend estimates discussed in the text are consistent with a linear extrapolation of these trends

freight traffic occurred by rail in 1999.[4]

Many scenarios have been developed to predict the future trends for surface traffic and the implications for energy use and emissions (cf. References 3 and 5). The OECD predicts in its EST study a further increase in road traffic of 23–76% by 2030.[3] The upper limit is based on a business as usual scenario with further rapid growth in developing countries and in Newly Industrialized Economies, an increase in transition countries and a stabilization in the OECD. As the world road sector performance grew by almost 100% in the last 30 years, the assumption of an increase of only 76% for the next 30 years already appears optimistic. The lower limit is based on a scenario with an environmentally sustainable transport system, implementing world-wide measures to reduce transportation demands, increase the efficiency, and promote the shift to collective transport modes and rail-based systems. Nevertheless, an increase of more than 20% transport performance is foreseen in the road sector, due to the strong development needs in many countries, mainly in Asia.

Traffic has already become a problem in several areas of the world, in particular in the urban agglomeration areas of industrial and developing countries. The high population density in urban areas of China and India will limit the future growth of road traffic in these regions. Today's traffic is largely dependent on mineral oil, and uses these resources much faster than they are replenished. Hence, surface traffic in its present form is clearly not sustainable.

[5] J. Dargay, and D. Gately, *Energy Policy*, 1997, **25**, 1121.

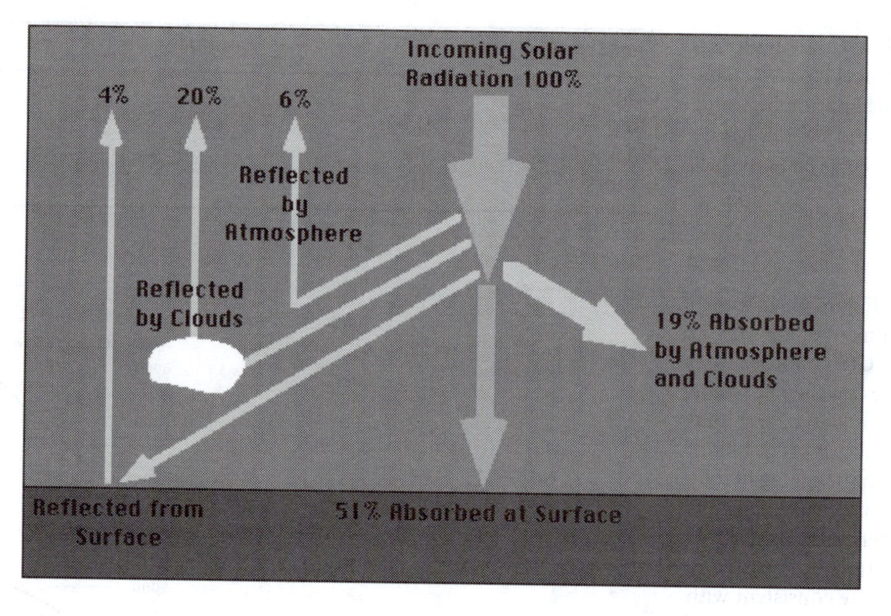

Figure 5 Schematic illustration of the atmospheric radiative balance

2 Mechanisms of Climate Impact

Surface traffic impacts the climate system by releasing large amounts of greenhouse gases (in particular carbon dioxide, CO_2) and other trace substances that react in the atmosphere to form ozone and aerosols, which then intercept the solar and terrestrial radiation and lead to additional warming of the Earth's atmosphere. Furthermore, some emissions from motorized traffic (notably nitrogen oxides, NO_x, and carbon monoxide, CO) influence the oxidizing power of the atmosphere, which controls the lifetime of greenhouse gases and thus also affects the climate forcing.

The weather is driven by radiation from the sun. Solar radiation is either returned to space, by reflection and scattering in the atmosphere and at the ground, or it is absorbed and radiated back towards space as thermal (long-wave) radiation (Figure 5). 36% of the incoming solar radiation (at the top of the atmosphere) is reflected and of the remaining 64% about one quarter is absorbed in the atmosphere and the rest at the Earth's surface. Spatial and temporal variations in the energy balance drive atmospheric motions. Besides the geometric factor (the equator receives 2.4 times as much energy as the poles), such variations are controlled by the land–sea distribution, the soil type and vegetation cover, clouds and by the chemical composition of the atmosphere (greenhouse gases and aerosol particles). Greenhouse gases absorb and re-radiate the outgoing thermal radiation, effectively storing some of the heat in the atmosphere, thus producing a net warming of the surface. This process is called the greenhouse effect. Water vapour is the most important greenhouse gas and it is highly variable. Other greenhouse gases include carbon dioxide (CO_2), methane (CH_4), nitrous oxide (N_2O), ozone (O_3), CFCs, and sulfur hexafluoride (SF_6). Without the greenhouse gases, the temperature of the earth would be determined by the

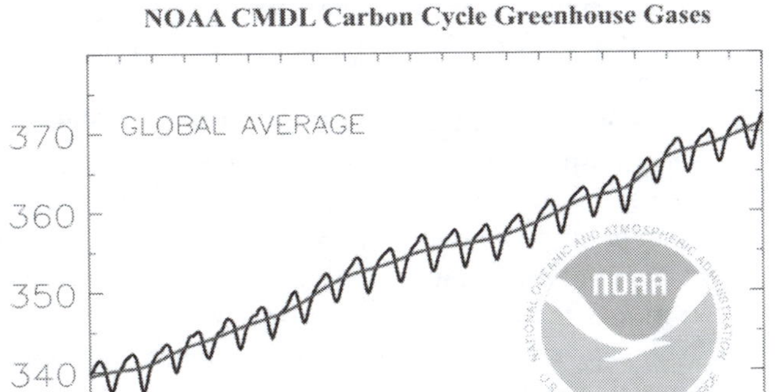

Figure 6 Global mean observed CO_2 concentration trend and CO_2 growth rate (from http://www.cmdl.noaa.gov/ccgg); due to the biogeochemical coupling of CO_2 between the atmosphere, ocean, and terrestrial vegetation, the contribution of surface traffic to the concentration rise can only be assessed with complex numerical models

amount of incoming solar radiation that reaches and heats the Earth's surface. To a good approximation the earth emits infrared radiation (heat) like a black body. According to the Stefan–Boltzmann law ($R = \sigma T^4$, where R is the amount of radiation emitted, σ is the Stefan–Boltzmann constant 5.6705×10^{-8} W m^{-2} K^{-4}, and T is the temperature of the surface) the global annual mean temperature would be -18°C without greenhouse gases. Measurements indicate a mean temperature of about $+15$°C, thus there is more than 30°C heating because of the greenhouse effect. The sources and sinks of greenhouse gases are closely linked to nutrient cycles and physical uptake in the ocean and the terrestrial biosphere. In view of these complex interactions, the impact of traffic emissions on the temporal evolution of greenhouse gas concentration can only be assessed by complex models (Figure 6).

The impact of the anthropogenic increase of greenhouse gas concentrations on the atmosphere is often described with a global mean quantity called 'radiative forcing'. This value denotes the difference in the net solar radiation at the top of the atmosphere with and without the greenhouse gas present (see Section 4). At the current concentration levels of greenhouse gases, the excess CO_2 (*i.e.* the amount emitted and stored in the atmosphere since the beginning of the industrialization) contributes about 1.5 W m^{-2} to the radiative forcing, methane (CH_4), N_2O, halocarbons, and tropospheric ozone contribute about 0.47, 0.14,

0.28, and 0.35 W m^{-2}, respectively.[6] In addition to the 'direct' greenhouse gases, species emitted from fossil fuel combustion engines also contribute to the 'indirect greenhouse effect'. Nitrogen oxides (NO_x), and carbon monoxide (CO) in particular act together in the formation of tropospheric ozone, and they influence the concentration of the hydroxyl radical ($^{\cdot}OH$), which is the most important reactant for atmospheric pollutants and greenhouse gases and thus determines the 'oxidizing power' of the atmosphere. At current concentration levels, CO acts as the main sink of tropospheric OH. NO_x plays a special role in determining the OH concentration, because it couples different chemical cycles. At low NO_x concentrations, increasing NO_x leads to higher $^{\cdot}OH$ levels; however, at typical urban NO_x concentration levels this relationship is reversed. The most significant consequence of this feedback is the dependence of the methane lifetime (and thus the methane concentration) on the $^{\cdot}OH$ concentration, and thus on the ozone precursor emissions.

Aerosol particles are released into the atmosphere as particles (primary particles) like dust, sea-salt or soot, or they are formed in the atmosphere by condensation of vapour (secondary particles), like nitrate, sulfate or organics. Aerosol particles affect the climate system by scattering and absorbing solar radiation. This leads to a cooling of the surface because the incoming solar radiation is reduced, and to a warming of the atmospheric layers carrying the absorbing aerosol (direct aerosol effect). Black carbon is a strong absorber of solar radiation; weaker absorption is caused by nitrate and organic particles. Sulfate does not absorb in the short-wave range. Furthermore, particles serve as cloud condensation and ice nuclei and impact the cloud's physical and optical properties (indirect aerosol effect). An increase in the aerosol concentration at the cloud base results in the formation of numerous smaller droplets compared with an unperturbed cloud. This makes the cloud brighter and results in a higher cloud albedo, *i.e.* the fraction of solar radiation scattered back to space is increased.[7] The production of more and smaller droplets also tends to decrease the efficiency with which precipitation is formed and thus prolongs the cloud's lifetime. Albrecht[8] proposed that a decrease in drizzle production within such clouds could increase both the cloud liquid-water content and the fractional cloudiness. For absorbing aerosols, this effect competes with another kind of indirect effect: Ackerman *et al.*[9] report observations that the heating of the boundary layer by absorbing aerosols may influence the formation of marine stratocumulus clouds by stabilizing the layer and by reducing relative humidity. However, simulations with a general circulation model[10] indicate that, while this effect may reduce cloudiness in some highly polluted locations, the increase in cloud albedo and lifetime is still predicted to dominate on the global scale. The effects discussed so far apply to low level liquid-water clouds, but aerosols also affect cirrus cloud properties. However, far

[6] V. Ramaswamy, O. Boucher, J. Haigh, D. Hauglustaine, J. Haywood, G. Myhre, T. Nakajima, G. Y. Shi and S. Solomon, *IPCC Third Assessment Report*, Cambridge University Press, Cambridge, 2001, p. 350.

[7] S. Twomey, *J. Atmos. Sci.*, 1974, **34**, 1149.

[8] B. Albrecht, *Science*, 1989, **245**, 1227.

[9] A. S. Ackerman, O. B. Toon, D. E. Stevens, A. J. Heymsfield, V. Ramanatha and E. J. Welton, *Science*, 2000, **288**, 1042.

[10] U. Lohmann and J. Feichter, *Geophys. Res. Lett.*, 2001, **28**, 159.

Table 5 Road traffic and total anthropogenic emissions of NO_x, CO, NMVOC, CH_4, and SO_2 for 1995 (EDGAR 3.2)[11]: note that 'anthropogenic' includes biomass burning emissions, which are very uncertain and rather significant, especially for CO, less so for NO_x, and probably not for the other species

Species	*Road traffic emissions*	*Total surface traffic emissions*[a]	*Total anthropogenic emissions*
NO_x ($TgNO_2$ yr^{-1})	28	41	110
CO (TgCO yr^{-1}	200	220	846
NMVOC (TgVOC yr^{-1})	33	40	153
CH_4 ($TgCH_4$ yr^{-1})	0	1	302
SO_2 ($TgSO_2$ yr^{-1})	4	5	142

[a]Including road traffic, non-road surface traffic and international shipping.

less is known about these effects, so that one cannot even determine if they lead to a net warming or cooling.

Aerosols can also influence the dynamical processes in the atmosphere that drive cloud formation and development. Aerosol-induced reduction of incoming solar radiation affects the energy balance at the ground, the surface heating, thermal emission, sensible and latent heat fluxes with subsequent changes in turbulent water vapour transport, cloud formation and evaporation rates.[12,13] Reduced evaporation is balanced by less precipitation. Unlike greenhouse gases, many of which are long lived and therefore well mixed in the troposphere, aerosols have a short atmospheric residence time (due to efficient washout by precipitation), resulting in substantial spatial and temporal variability.

In addition, there are feedback processes between aerosols and the indirect greenhouse gas effect, which are only now being addressed in the climate research community. For example, the absorption and scattering of aerosols in the UV and visible spectral range influence the photodissociation rates of gas-phase molecules, which may have an important effect on the global mean $^{\cdot}OH$ concentration. Conversely, $^{\cdot}OH$ plays an important role in the oxidation of sulfate precursors and may thus affect the formation rate of new particles. Furthermore, the chemical composition of aerosol particles determines their acidity, which also influences the hygroscopicity (uptake of water vapour), and therefore has consequences for the size and the physical properties of the particle.

Mainly due to anthropogenic emissions, the atmospheric volume mixing ratio of carbon dioxide has increased from 280 to 350 ppm, that of methane from 0.7 to 1.7 ppm, that of nitrous oxide from 0.28 to 0.31 ppm, and the tropospheric ozone concentrations have nearly doubled between 1860 and 1990.[14] Climate model simulations show that this increase in emissions is consistent with the observed warming of about 0.6°C at the surface observed over the past four decades and that this warming cannot be explained by natural variations alone.[14] Not only

[11] J. G. J. Olivier and J. J. M. Berdowski, Global emission sources and sinks, in J. Berdowski, R. Guicherit and B. J. Heij (eds.), *The Climate System*, A. A. Balkema and Suets & Zeitlinger, Lisse, The Netherlands, 2001, 33.

[12] E. Roeckner, L. Bengtsson, J. Feichter, J. Lelieveld and H. Rodhe, *J. Climate*, 1999, **12**, 3004.

[13] V. Ramanathan, P. J. Crutzen, J. T. Kiehl, D. Rosenfeld, *Science*, 2001, **294**, 2119.

[14] J. T. Houghton, Y. Ding, D. J. Griggs, M. Noguer, P. J. van der Linden, X. Dai, K. Maskell and C. A. Johnson, *IPCC Third Assessment Report*, Cambridge University Press, Cambridge, 2001, p. 1.

[15] J. E. Penner, M. Andreae, H. Annegarn, L. Barrie, J. Feichter, D. Hegg, A. Jayaraman, R. Leaitch, D. Murphy, J. Nganga and G. Pitari, Aerosols, their Direct and Indirect Effects, *IPCC Third Assessment Report*, Cambridge University Press, Cambridge, 2001, p. 289.

Table 6 CO_2 emissions in $TgCO_2$ for 1995 for road transport, non-road surface transport and international shipping (EDGAR 3.2[11], Olivier *et al.*, 2003)

Region	*Road transport*	*Non-road surface transport*	*International shipping*	*Total anthropogenic*
Canada	110.2	22.7	2.0	508.8
USA	1342.8	84.7	86.0	5576.2
OECD Europe	747.0	32.4	115.3	3719.0
Oceania	71.3	7.6	3.9	371.2
Japan	217.7	13.1	19.4	1284.6
Eastern Europe	60.1	4.3	1.7	1001.0
Former USSR	75.1	108.9	20.1	3096.5
Latin America	344.3	26.8	27.3	2013.7
Africa	111.4	3.7	26.3	1244.3
Middle East	209.2	1.5	48.0	1153.3
South Asia	124.0	9.3	2.8	1191.6
East Asia	210.2	63.7	37.3	4535.8
South East Asia	137.4	6.9	39.4	1242.8
Total	3760.6	385.7	429.3	26938.7

greenhouse gas concentrations but also the amount of tropospheric aerosols has increased substantially over the past 150 years as a consequence of fossil fuel use and biomass burning. About 70% of the sulfate aerosol precursor sulfur dioxide is anthropogenic, as is 74% of nitrate, 51% of organic matter and almost 100% of black carbon.[15]

3 Emissions from Surface Traffic

To assess the contribution of surface traffic to the emissions of greenhouse gases, ozone precursors, and aerosol particles it is important to establish the reference frame for the comparison. Several atmospheric species have large anthropogenic as well as natural sources, and some compounds (*e.g.* CO) are formed to a large degree in atmospheric chemical reactions between precursor species, which may potentially result from traffic emissions as well. In a summary statement, the International Energy Agency[16] notes that surface traffic accounted for about 11% of all global greenhouse gas emissions for 1996. Relative to the total direct anthropogenic emissions, IPCC (Reference 13 p. 257 ff) lists a contribution from fossil fuel combustion of 80% for NO_x, and of 41% for CO. Compared with the estimates of the total source of these compounds, these fractions decrease to 63, and 23%, respectively.

In industrialized countries, emissions from surface transport are significant. In Germany in 1999, the fractional contribution of surface transport to the total emissions was 21% for carbon dioxide, 61% for NO_x, 52% for CO, 23% for NMVOC, and 20% for aerosols.[17] In a global average sense, these numbers are somewhat lower due to the large contributions from, mainly, biomass burning

[16] IEA, *CO_2 Emissions from Fuel Combustion 1971–2000* (2002 Edn.) International Energy Agency, Paris.
[17] UBA—Umweltbundesamt, Emissionen des Verkehrs. Berlin (www.env-it.de/umweltdaten), 2003.
[18] UNEP—United Nations Environmental Programme, *Global Environmental Outlook 3 (GEO3)*, Earthscan, London, 2003.

Table 7 Changes in abundance (greenhouse gases) or emissions (aerosol) of radiative active species released by surface traffic and their radiative forcing

	Change in abundance/ emissions	*Normalised forcing per abundance or annual emission*	*Radiative forcing ($W\,m^{-2}$)*
Carbon dioxide	17.4 ppm	16.78×10^{-3} W m^{-2}/ppm[a]	+0.29
Methane	−100.5 ppb	$0.459.10^{-3}$ W m^{-2}/ppb[a]	−0.046
Ozone	0.8224-1.234 DU	0.033-0.056 W m^{-2}/DU[a]	0.034 to 0.051
Sulfate (direct)	0.01 g(SO_2)m^{-2} yr^{-1}	3[a]–7[b] W yr/g (SO_2)	−0.03 to −0.07
Soot (direct and indirect, only car traffic)	52.9×10^{-6} g^{-2}	1210[c]–3100[d] W g^{-1}	0.064 to 0.16
Aerosol effect on clouds (only ships)			−0.11[e]
Total			0.16 to 0.32

References: [a]6; [b]22; [c]24; [d]26 and [e]31.

and agricultural emissions in tropical regions (Table 5). In general, energy supply and transport are the two main activities responsible for the overall growth in global greenhouse gas emissions. Households and agriculture have a far slower growth speed, and industry shows a world-wide reduction trend for greenhouse gases.[4]

Looking at the spatial distribution of CO_2 emissions, the main transport-related source is in North America (Reference 17, Table 6). OECD countries are responsible for more than two-third of road transport CO_2 emissions. International shipping is responsible for less than 10% of the CO_2 emissions from surface traffic.

Emission estimates are, notably, still rather uncertain, partly due to incoherent reporting on the national level, and partly due to the lack of measurements to relate fuel consumption and the composition of the vehicle fleet to emission factors. The situation is particularily dire for aerosol particles and for specific hydrocarbon compounds, because they are difficult to measure quantitatively and meaningful reporting standards still have to be developed. Nevertheless, Tables 5 and 6 clearly show that surface traffic plays a major role in polluting the atmosphere on a global scale.

4 Climatic Impacts of Traffic Emissions

So far there are no published climate change studies based on advanced climate model simulations that focus on the impact of surface traffic. Therefore, we will make use from existing estimates of the radiative forcing due to different external perturbations to derive a rough estimate of the possible climate impact. The large uncertainties of many of the terms involved do not warrant a too detailed treatment of the radiative forcing, therefore we assume that the relationship between emissions and radiative forcing is linear. Thus, a given percent increase in traffic emissions will translate into the same percentage of the radiative forcing attributed to traffic. Radiative forcing (RF) is an externally imposed perturbation of the radiative energy budget of the Earth's climate system. RF is defined here as the sum of both long- and short-wave changes in the radiation budget. RF is calculated by atmospheric models applying the radiative calculations twice, once with the perturbation and once without. The meteorology is kept constant and the RF is then the difference of the radiative fluxes at the top of the troposphere. The climate sensitivity λ is the response of the near-surface temperature to a specific forcing and is defined as

$$\Delta T_s / \Delta RF = \lambda$$

Values of λ are model dependent but within each model it is, to a first order, nearly invariant to a wide range of different forcings.[19] Values of λ range between 0.5 and 1 K W^{-1} m^2.[6] By adding the different radiative forcings and assuming an invariant climate sensitivity, we obtain a rough estimate of the surface temperature response due to surface traffic emissions.

To estimate the forcing of the greenhouse gases CO_2 and CH_4 we assume the same forcing per change in abundance as given in IPCC.[14] For CO_2 this is 16.78 $\times$ 10^{-3} W m^{-2} per 1 ppm change in CO_2 concentration, and for CH_4 it is 0.459 W m^{-2} per 1 ppm change in CH_4 concentration. The climate impact of the short-lived greenhouse gas ozone is estimated based on published normalized forcing estimates (W m^{-2} per Dobson unit). Model calculated values of tropospheric ozone changes are in the range 0.033 to 0.056 W m^{-2} DU^{-1} with a mean value of 0.042 W m^{-2} DU^{-1}.[6,19] Estimates of RF based on this mean value are given in Table 7.

Granier and Brasseur[21] have used a global chemistry transport model to investigate the impact of surface traffic on tropospheric ozone. They found a summer-time increase in ground level ozone concentration between 5 and 15% over most of the Northern Hemisphere when comparing two simulations with and without emissions from road traffic. This is consistent with model results from Schultz *et al.*,[34] who reduced the global emissions of NO_x and CO by 50% in a study investigating the potential impacts of hydrogen-powered vehicles on climate and air pollution. Translating the relative ozone increase into a radiative forcing estimate, we find a positive forcing of about 0.04 W m^{-2} caused by surface traffic emissions (Table 7). Both studies find an increase in the oxidizing power of the atmosphere (the $^{\cdot}OH$ concentration) due to NO_x emissions from traffic. This leads to a decrease in the methane lifetime of about 3%, which would reduce the CH_4 concentration and thus the CH_4 forcing accordingly.

Surface traffic emissions also contribute to the atmospheric aerosol load. NO_x and SO_2 in the atmosphere are chemically transformed into nitrate (NO_3^-) and sulfate (SO_4^{2-}), respectively. These condensable vapours form particles or condense on ambient pre-existing particles. Both nitrate and sulfate mix with ammonium (NH_4^+) and water (ternary mixture). Present day (year 2000) global and annual average anthropogenic direct radiative forcing is estimated by Adams *et al.*[22] to

[19] J. Hansen, M. Sato and R. Ruedy, *J. Geophys. Res.*, 1997, **102**, 6831.

[20] D. A. Hauglustaine and G. P. Brasseur, *J. Geophys. Res.*, 2001, **106**, 32 337.

[21] C. Granier and G. P. Brasseur, *Geophys. Res. Lett.*, 2003, **30**, article no. 1086, doi 10.1029/2002GL 015972.

[22] P. J. Adams, J. H. Seinfeld, D. Koch, L. Mickley, D. Jacob, *J. Geophys. Res.*, 2001, **106**, 1097, doi

be -0.95 and -0.19 W m^{-2} for sulfate and nitrate, respectively. Their normalized direct RF for sulfate, defined as RF per column aerosol mass, is at -290 W g^{-1} on the high side of current estimates which range between -110 and -250 W g^{-1}.[6]. The normalized direct RF for nitrate was estimated by Adams *et al.*[22] at 255 Wg^{-1}. So far there is no other estimate of the forcing from nitrate available. If we relate the direct RF to the source strength of the aerosol precursors sulfur dioxide and nitrogen oxide, the normalized direct RF, defined as RF in W m^{-2} per emission flux in g(S) m^{-2} yr^{-1}, is 7.0 W yr g^{-1} for sulfate and 2.1 W yr g^{-1} for nitrate.

Diesel powered engines emit a considerable amount of fine particulates, which are released directly as smoke and diesel soot. These particles consist of black carbon (BC) internally mixed with organic matter (OM). Calculations of the RF were performed for OC and OM from fossil fuel use emissions. The normalized direct RF of different model estimates ranges between 1200 and 3000 W g^{-1}. This large scatter is due to different assumptions about the degree of mixing with sulfate. Recently, we performed simulations with a coupled ocean–atmosphere model including some aerosol physics and sulfur chemistry[23] to study the climatic impact due to the direct and the indirect effect of BC emissions from surface traffic[24] The amount of BC emitted by diesel powered cars was assumed to be 1.8 Tg carbon per year.[25] Compared with BC emissions from total fossil fuel use, the contribution of cars corresponds to 12%. The amount of BC in the atmosphere from car emissions was calculated to be 27 Gg C as a global and annual average. The overall effect of BC is a warming in the polluted regions where surface temperatures could increase by 0.2 to 0.5 K. However, the global and annual mean average temperature difference between a simulation with, and one without, BC emissions from cars is negligible ($\Delta T = 0.04$ K). This finding is in contrast to a recent study by Jacobson[26] who points out that warming due to BC and OM emissions (0.35 K) from fossil fuel use amounts to one-third of the warming due to the anthropogenic CO_2 increase. Jacobson[26] reports a normalized direct RF of fossil fuel use BC of 3100 W g^{-1} (Reference 27) whereas this normalized forcing in our simulation is 1210 W g^{-1}.[28] Different assumptions about the physical and optical properties and a longer atmospheric residence time of the BC particles due to less efficient removal of particles by precipitation results in a higher aerosol load and explains the differences between Jacobson's and Hendricks and Feichter's estimate. The two estimates roughly represent the uncertainty of the climate impact from carbonaceous aerosols.

While sulfur dioxide emissions from road traffic are rather low, ocean-going ships significantly perturb the sulfur cycle in remote marine regions. Estimates of the sulfur emissions from ships indicate that in many regions these emissions are of the same order of magnitude as the release of dimethyl sulfide (DMS) from the

10.1029/20003D 900512.

23 J. Feichter, E. Roeckner, U. Lohmann and B. Liepert, *J. Climate*, 2003, **17**(2), 2384.

24 J. Hendricks and J. Feichter, in preparation.

25 I. Köhler, M. Dameris, I. Ackermann and H. Hass, *J. Geophys. Res.*, 2001, **106**, 17 997.

26 M. Z. Jacobson, *J. Geophys. Res.*, 2002, **107**, 4410.

27 M. Z. Jacobson, *Nature*, 2001, **409**, 695.

28 W. F. Cooke, C. Liousse, H. Cachier and J. Feichter, *J. Geophys. Res.*, 1999, **104**, 22 137.

marine biosphere, which constitutes the most important non-anthropogenic source of sulfur.[29,30] Globally the sulfur emitted by ships corresponds to roughly 20% of the biogenic emissions from the oceans. Remote marine regions are very susceptible to the indirect aerosol effect due to the low air pollution, the low concentration of cloud condensation nuclei and the high water content of low level clouds. Capaldo *et al.*[31] estimate the climatic impact of ship emissions *via* the cloud albedo effect to be 0.11 W m^{-2}, with a stronger effect in the Northern (0.16 W m^{-2}) than in the Southern Hemisphere (0.06 W m^{-2}). This forcing is about 10% of the total aerosol forcing from all anthropogenic emissions.[10] However, SO_2 from ocean-going ships accounts for only 3–4% of the total emissions of these species from the burning of fossil fuels.[32] This emphasizes the sensitivity of clean air marine regions against air pollution from ships. Besides SO_2, ships also release large amounts of soot and NO_x. However, estimates of the source strength of soot emissions from ships are still unavailable. The impact of shipborne NO_x emissions has been investigated in different modelling studies,[30,33] which arrive at contradictory conclusions regarding their impact on tropospheric ozone. Given the magnitude of ship NO_x emissions relative to emissions from road traffic, it is clear, however, that the NO_x related changes in ozone will not have a significant climate impact.

The sum of all radiative forcing terms in Table 7 amounts to 0.16 to 0.32 Wm^{-2}. Given that the climate sensitivity is between 0.5 and 1 K W^{-1} m^2 our estimate of the surface temperature response due to surface traffic emissions ranges between 0.08 and 0.3 K. The upper bound of this estimate is similar in magnitude to the response due to the anthropogenic increase of tropospheric ozone. While this appears a relatively modest effect on the global scale, there might be much larger perturbations regionally. Even if there were a cancellation of positive and negative forcings in the global mean, regional responses may be different from zero.

5 Future Developments and Possible Mitigation Options

For industrialized countries the challenge lies in the preservation of the current mobility levels, while at the same time the dependence on fossil fuels must be reduced to keep the vehicle and ship fleets running, even if oil reserves will eventually be depleted. The current emissions from fossil fuel use in mobile engines constitute a large fraction of CO_2 and ozone precursor emissions, which have to be reduced to meet the Kyoto protocol as well as reasonable air quality standards. If a sustainable atmosphere is to be achieved, traffic emissions must be practically eliminated. Developing countries bear a much smaller share of current CO_2 emissions, but they often suffer from severe air quality problems (see http://www.who.int/inf-fs/en/fact187.html), which will be aggravated if the current

[29] J. J.Corbett and P. Fischbeck, *Science*, 1997, **278**, 823.

[30] J. J. Corbett, P. S. Fischbeck and S. N. Pandis, *J. Geophys. Res.*, 1999, **104**, 345

[31] K. Capaldo, J. J. Corbett, P. Kasibhatla, P. Fischbeck and S. N. Pandis, 1999, *Nature*, **400**, 743.

[32] P. Sinha, P. V. Hobbs, R. J. Yokelson, T. J. Christian, T. W. Kirchstetter and R. Bruintjes, *Atmos. Environ.*, 2003, **37**, 2139.

[33] M. G. Lawrence, P. J. Crutzen, *Nature*, 1999, **402**, 167.

increase in oil-powered motor vehicles continues.

Several mitigation options have been proposed, ranging from fuel-saving vehicles (the so-called '3-litre car') to a radical shift towards alternative fuel options, such as hydrogen-powered fuel cells. In the following, we concentrate on the available options for road traffic, as this constitutes the dominant fraction of surface transport at present. However, we note that, in the medium-term, ship emissions may need to be controlled as well.

Three main strategies are currently in use, mainly in industrialized countries, to enhance the environmental performance of the surface traffic and to mitigate the climate impact through reducing the emissions of carbon dioxide and other pollutants:

- Increase energy efficiency and implement new filter technologies,
- facilitate the use of regenerative propulsion technologies and
- reduce overall transport demand.

A wide range of economic instruments like tax, pricing or emission trading are in use or under development, in order to implement these strategic objectives. E. Calthrop and S. Proosr, Environmental pricing in transport, in D. A. Henster and K. J. Button (eds.), *Handbook of Transport and the Environment*, Elsevier, Amsterdam, 2003, 529.

Even if a net reduction of surface traffic is a main objective for a multitude of municipalities, especially in the central areas of the main cities and in peak times or during pollution episodes, this objective is not a main target of any coordinated international activity.

Because policies for environmentally friendly transport have been pursued for many years in many countries and managed by many international organizations such as the EU, OECD or ASEAN, or diverse UN bodies such as the Global Initiative on Transport Emissions, an overall independent evaluation of these measures could be performed. However, no such independent scientific evaluation is currently available.

Therefore, even a summary view of the main achievements faces a lack of empirical data on emissions reductions related to different measures and their specific impacts. For example, the introduction of a new metro system (sky train) in Bangkok has been recognized as a very powerful measure for reducing traffic jams and the level of pollution in this agglomeration of 10 million inhabitants. Parallel to the sky train, old vehicles and, especially, diesel buses, causing more than 95% of the aerosol emissions, have been replaced by the municipality. Intuitively, the overall effect should be a significant reduction in greenhouse gas and ozone precursor emissions. However, no scientific evaluation has been published comparing the overall amount of pollutants emitted by road traffic before and after the introduction of the measures. Data on vehicle movements should be available, thanks to a sophisticated monitoring system.

Efficient use of fossil fuels is a strategic activity with broad official support, because it leads to a higher economic value with less input of primary resources. Regarding oil depletion, efficiency is also the first choice strategy for many industries and governments, but it has the negative effect of confirming the technological 'lock-in' based on diesel and gasoline motorization, oil distribution systems and the petroleum production chain. For emission reduction, efficient technologies have mainly a positive impact on CO_2, but less so on other pollutants. Therefore, there is a need for additional filter or clean motor and vehicle technologies to reduce aerosols, NO_x or CO, and further reduce the very high energy losses (experts estimate that a mean of more than 80% of end energy is lost in conversion in the mobility system). Numerous efficient technologies have a proven positive impact on fuel consumption and emissions, while preserving a constant quality of mobility services.

'Regenerative technologies' are considered by specialists as the key for a sustainable transport but, at the present time, no technology is positioned to achieve a clear market dominance. It is, therefore, likely that the transition towards more regenerative technologies (such as hydrogen propulsion, electric vehicles, biofuels *etc.*) will have to take into account a broad range of different utilizations and propulsion systems, not in a matter of concurrence, but in a complementary manner. Of course the primary energy source should be produced in a sustainable way, as is the case for wind or solar electricity, geothermal heat, and for some biomass or bio-waste energy. Otherwise, the paradoxical situation of having increased climate forcing with seemingly 'clean' technologies might arise.[34] Additionally, the entire distribution and mobility system could be more sustainable.

Under the concept 'fuel switch', the actual use of bio-diesel, methanol, natural gas or hydrogen have, by far, different consequences for the atmosphere and emission reduction, depending on the conditions of production, distribution and use, as the results from the FANTASIE project of the EU have shown. Despite remaining problems, these fuels are all considered to be feasible bridging technologies, facilitating the shift away from oil. One of the newest and promising combinations is the geothermal/hydrogen system currently tested in Iceland, a technological option with very low emissions considering the entire mobility, energetic and industrial system involved, including the industry producing the electric vehicles.

6 Conclusions

Surface traffic contributes significantly to the anthropogenic emissions of greenhouse gases, ozone precursors, and aerosols. The impact of traffic emissions on the climate system is very poorly investigated, which is in part due to a lack of reliable emission estimates and targeted measurements linking traffic activity to emission factors. Using a simplified approach and currently accepted values for radiative forcing terms from the literature, we find that surface traffic is responsible for about 12% of the total present greenhouse gas forcing of 2.4 W m^{-2}.[14] Most of this forcing is due to the CO_2 emissions from fossil fuel combustion engines; other factors tend to cancel out.

Investigating the climate impact of future traffic scenarios is difficult, because of the large parameter range for future technological options and the close coupling between various forcing mechanisms. Comprehensive models simulating

[34] M. G. Schultz, T. Diehl, G. P. Brasseur and W. Zittel, *Science*, 2003, **302**, 624.

not only the physical climate system, but also the chemistry of trace gases and aerosol particles are currently under development at several institutions and they represent the only way to reliably assess the potential climate perturbations in a future world.

If no measures are taken and the current growth rate of surface traffic continues, then transport will soon become a major cause for climate change. Mitigation strategies must take into account the knowledge gap between industrial and developing countries, and it must be ensured that new technologies are directly implemented in developing countries, because they experience the fastest growth in traffic related emissions. A sustainable development of emerging economies can only be achieved when the historical mistakes of the Northern Hemisphere countries are not repeated.

Unfortunately, the spatial structures of cities and urbanized regions in industrialized countries as well as the perception of mobility as a positive asset are leading to an increasing demand for mobility, separating the place of work from the housing or commercial quarters, and therefore contributing to high energy use and the concentration on road transport infrastructure. New concepts in the fields of demand-side management, city transport planning or classical rail and metro promotion and subvention have all shown substantial effects, leading to far less emission and far better living conditions in innovative municipalities such as Mexico City, Zürich and Bangkok. However, the overall effects of all these local measures are, on the global scale, by far overcompensated by the fast growing world mobility demand.

Human Health Implications of Air Pollution Emissions from Transport

STEPHEN B. THOMAS AND ROY M. HARRISON

1 Sources and Emission Inventories of the Transport Sector

Transport has long been a significant source of air pollution and, consequently, there is considerable concern over the effects of transport emissions on human health. Urban road traffic is a principal source of transport emissions and much of the scientific review information describing the associated health effects have concentrated upon emissions resulting from this mode of transport.[1]

In an extensive review of the transport sector as a source of air pollution, Colville *et al.*[1] consider the effects of various modes of transport. However, of the three transport modes presented (road traffic, aircraft, and shipping) only the impacts of road traffic on human health are addressed. While the impacts of road transport on human health may be currently receiving considerable attention, there are gaps in the understanding and quantification of the health impacts of all of the transport modes, including, in addition to road transport, shipping, aircraft, and rail.

All modes of transport emit air pollution. For most of the transport sector these emissions generally result from the combustion of liquid fossil fuels. However, in contrast to stationary sources utilizing fossil fuel combustion, for mobile sources the combustion is incomplete, and a small fraction of the fuel is oxidized to carbon monoxide (CO) with some volatile hydrocarbons, and carbonaceous particles are also emitted.

In addition, nearly all fuels contain some impurities. In the combustion process, sulfur is oxidized mostly to sulfur dioxide (SO_2), and sometimes to sulfate which can assist in the formation of particles in the exhaust. Other impurities (*e.g.* vanadium in oil) do not burn or have combustion products that contribute further to particle formation. In Africa and Asia, organic lead compounds that are still employed in fuels, to prevent premature combustion, also form particles in the exhaust. Finally, at the high combustion temperatures required

[1] R. Colville, E. Hutchinson, J. Mindell and R. Warren, *Atmos. Environ.*, 2001, **35**, 1537–1565.

Issues in Environmental Science and Technology, No. 20
Transport and the Environment

for most modes of transport, atmospheric nitrogen (N_2) is oxidized to nitric oxide (NO), and small quantities of nitrogen dioxide (NO_2). Further nitrogen dioxide is formed from atmospheric oxidation of NO.

For most transport modes the emissions occur in the location of the transport use. An exception is air pollution from the operation of electric railways and the road vehicles that are powered by electricity. For these modes, SO_2 and other emissions from the generation of electricity occur remote from the place of use.

The *key primary emissions* from the transport sector that may have an impact on air quality are therefore:

- CO
- Oxides of nitrogen (NO_x)
- Hydrocarbons (VOCs)
- SO_2
- Particulate matter (PM)
- Air toxics (including organic compounds and metals)

Since there are various sources of these atmospheric pollutants it is not possible to measure emissions from all of the individual sources or, even, from all the different source types. In practice, atmospheric emissions are estimated on the basis of measurements made at selected or representative samples of the sources and source types.

Emission estimates are modelled on the basis of (at least) two variables, for example:

- an activity statistic and a typical average emission factor for the activity, or
- an emission measurement over a period of time and the number of such periods emissions occurred in the required estimation period.

Emission inventories are then developed based upon the collection of data from these estimates. These inventories are helpful in estimating the contribution of transport to air pollution emissions compared with another activity, or the relative contribution of alternative modes of transport. For example, the emission inventories for the UK, EU, US, and various European countries are provided in Table 1, illustrating the contribution of road transport (RT) and transport modes other than road transport (OT) relative to the total emissions from all activities. Misclassification of the various modes of transport other than road transport is a commonly encountered problem that is highlighted by the lack of systematic data for the OT sector for many of the European countries. Apart from the 'on-road' vehicles (passenger cars, light duty vehicles, heavy duty vehicles, buses, two wheelers), internal combustion engines are used in many other modes of application and in some cases there is a risk of overlapping between source categories.

Inspection of the inventories presented reveals that RT is the most important contributor of CO in practically all of the inventories, and significant contributions of road transport to the emission levels of NO_x, and non methane volatile organic compounds (NMVOCs) are also a feature. Another feature of the data illustrated by the PM data for the UK is the increasing relative contribution (from 18 to 30%) of the road transport sector to particle emission levels with

Table 1 Contributions of road transport (RT) and other modes of transport (OT) to selected pollutant emissions by percentage of total emissions for UK (in 2000)[7], European Union members (EU 15 in 1999)[2], European Accession Countries (AC 9 in 1999)[2], US (in 1999)[3], and various European countries (in 1999)[4]

	CO		*NO_X*		*NMVOC*		*SO₂*		*PM₁₀*		*PM_{2.5}*		*PM₁*	
	RT	*OT*	*RT*	*OT*	*RT*	*OT*	*RT*	*OT*	*RT*	*OT*	*RT*	*OT*	*RT*	*OT*
UK	69	11	42	11	24	4	1	3	18	6	24	5	30	7
EU 15	57	7	45	18	31	6	3	4	28[a]	11[a]				
AC 9			37	12	37	5	2	1						
US	51	26	34	22	29	18	2	5	1.4[b]	2.2[b]	3.4[b]	6.1[b]		
Austria	24.2		40.8		9.9		7.1		12.8					
Belgium	53.3		48.8		30.3		3.3		12.2					
Germany	53.0		50.9	11.9	20.4		3.1		16.1					
Denmark	56.0		36.8		34.2	2.6	1.9	0.9	13.0					
Spain	53.8		39.9	15	15.2		1.5		16.1					
Finland	48.6		45.9		31.7	5			11.7					
France	41.5		51.4		25.5		4.9		11.3					
Italy	68.1		50.3		43.6		1.0		14.7					
Luxembourg	64.0		43.8	9–19	37.5		25.0		8.8					
Netherlands	60.5		41.8	6.5	36.0	5–12	4.9	1–3	14.7					
Sweden	57.5		44.9		21.8	1.6	1.9	2.6	13.9					

[a]Emissions of particulates assigned as primary and secondary fine particulates of which 12% are considered primary PM_{10}.
[b]Direct emissions only (*i.e.*, does not include fugitive dust).

Table 2 Contributions of emissions estimates for various pollutants (kt) in the US for 1999[3]

Source	SO_2	*CO*	NO_X	PM_{10}	$PM_{2.5}$	*NMVOC*
On-road vehicles total	*363*	*49989*	*8598*	*295*	*229*	*5297*
• LDV – gas	• 136	• 27187	• 2846	• 58	• 34	• 2870
• Motorcycles	• 0	• 195	• 13	• 0	• 0	• 42
• LD trucks – gas	• 91	• 16115	• 1638	• 36	• 22	• 1722
• HDV – gas	• 17	• 4262	• 459	• 12	• 8	• 375
• HDV – diesel		• 2217	• 3620	• 186	• 164	• 284
• LD trucks – diesel	• 118[a]	• 5	• 6	• 1	• 1	• 2
• LDV – diesel		• 8	• 8	• 1	• 1	• 3
Non-road engines and vehicles	936	25162	5515	458	411	3232
Aircraft	12	1002	175	38	27	183
Marine vessels	273	138	1007	46	40	34
Railroads	113	119	1204	30	27	49
Non-road other	3	883	235	2	2	0
Transportation total	*1299*	*75151*	*14105*	*753*	*640*	*8529*
Emissions total	*18867*	*97441*	*25393*	*20634*	*6773*	*18145*

[a] All diesel vehicles.

Table 3 Estimates of the 1990 emissions of EU countries for different vehicle categories as percentage of the EU totals for road transport[2]

Category	*CO*	*NO_X*	*NMVOC*	*PM*
Gasoline passenger cars	79.15	28.2	58.69	0
Diesel passenger cars	1.00	6.20	1.86	17.89
Gasoline LDV	9.57	3.84	7.72	0
Diesel LDV	1.17	7.65	1.54	21.65
Diesel HDV	3.75	47.07	16.8	55.83
Buses	0.43	6.70	1.18	5.39
Coaches	0.04	0.85	0.18	0.69
Moped	1.91	0.02	8.80	0
Motorcycle	3.69	0.20	3.86	0

decreasing particle size (for PM_{10} to PM_1)*. Adding to the concerns about the levels of PM attributed to road transport is that for this source the emissions are released at ground level close to human receptors. Hence there is less opportunity for the atmosphere to dilute the emissions. In addition, the emission concentrations may be increased by the fact that many roads have buildings alongside. The effect of such buildings is to shelter the road, reducing the wind speed at the source of emissions by as much as an order of magnitude relative to that on an open road.[5]

Table 2 provides the emission inventory estimating the contributions of all modes of transport categorized for the US for 1999. The most important source of CO is road transport and, in particular, petrol-driven vehicles, contributing about 50% of all CO emissions across the US. In areas of increased traffic congestion this figure may be as high as 95%.

Table 3 presents an inventory illustrating the relative contributions for categorized on-road vehicles for the EU. All road transport emits PM but diesel vehicles emit a greater mass of particulate matter per vehicle kilometre than petrol-engined vehicles. Emissions also arise from brake and tyre wear and from the re-entrainment of dust from the road surface. Emission estimates for the *re-suspension* of dust from roads are not included in the standard UN/ECE reporting format (and hence was not included in the original table reported by EU). However, estimates for *resuspension* indicate that, in 2000, 11% of UK PM_{10} emissions may have arisen from (re)suspension of surface deposits. Such estimates are, however, extremely uncertain.

Emission inventories may serve as a guide to atmospheric emissions but not necessarily to urban air quality. Where different sources have different heights of emissions or have special operational characteristics producing diurnal or seasonal variations the inventories are less informative. These factors are especially crucial for airborne particulate matter, with the result that the contributions from the different source categories to the primary emissions in many inventories do not reflect their respective contributions to urban air quality.[6]

* PM_{10} and PM_1 approximate to particles of diameter less than 10 and 1 μm respectively.

2 J. Goodwin and K. Mareckova, *Emissions of Atmospheric Pollutants in Europe*, 1990–1999, Topic Report 5/2002, 2002.

3 USEPA, *National Air Quality and Emissions Trends Report 1999*, www.epa.gov/oar/aqtrnd99, 1999.

4 European Environment Agency, *Emission Inventory Guidebook*, 2002.

5 L. Morawska, S. Thomas, D. Gilbert, C. Greenaway and E. Rijnders, *Atmos. Environ.*, 1999, **33**, 1261.

6 APEG (Airborne Particles Expert Group), *Source Apportionment of Airborne Particulate Matter in the United Kingdom*, 1999.

An evaluation of emission inventories with respect to measured air quality can provide a semi-quantitative verification of the emission estimates. This type of comparison has enabled the validity of the spatial distribution of these sources for a range of pollutants within the UK National Atmospheric Emissions Inventory to be tested.[7]

2 Estimates of Ground-level Concentrations of Air Pollutants Attributable to Transport Emissions

Numerical modelling of the health effects of transport-related emissions on local air quality requires spatially disaggregated emission inventories. Most of the investigations of these types of emissions have focussed mainly on road traffic in urban areas.[8–12] Examples of various case studies will be considered here to illustrate the issues involved in quantifying the contributions of transport-related emissions to the measured atmospheric concentrations of pollutants.

Estimates of Pollutant Concentrations Attributable to Road Traffic

Receptor modelling techniques have been extensively employed by the UK Airborne Particles Expert Group (APEG)[6] to estimate the contributions of road traffic to PM_{10} in 17 UK urban background sites and were further developed by Stedman *et al.*[13,14] for a site in Central London along with site specific projections back to the early 1990s and forward to 2010.

The concentrations of CO and NO_X were used as tracers of traffic emissions.[13,14] The air monitoring data were split into four seasons, and within each season data were extracted, for example, for CO and PM_{10} for twelve alternate hours of the day. For any given hour, the paired data for CO and PM_{10} for a given site at that time on each day in the appropriate season of the year were pooled and the relationship between the CO and PM_{10} concentration calculated using linear regression analysis. The researchers found highly significant correlations in summer and winter reflecting a relationship. Although virtually all emissions of CO within an urban area arise from road traffic, there was expected to be an external contribution to the urban concentration. Background CO concentrations were estimated as 0.14 ppm in the winter and 0.1 ppm in the summer from the Northern Hemisphere CO background according to the work of Derwent *et al.*[15] The regression equation between CO and PM_{10} was then used to estimate

[7] J. Goodwin, A. Salway, C. Dore, T. Murrells, N. Passant, J. Watterson, M. Hobson, K. Haigh, K. King, S. Pye, P. Coleman and C. Connolly, *UK Emissions of Air Pollutants 1970–2000*, 2002.
[8] J. Brandt, J. Christensen, L. Frohn and R. Berkowicz, *Phys. Chem. Earth*, 2003, **28**, 335–344.
[9] R. M. Harrison, A. Deacon, M. Jones and R. Appleby, *Atmos. Environ.*, 1997, **31**, 4103–4117.
[10] D. Briggs, C. de Hoogh, J. Gulliver, J. Wills, P. Elliott, S. Kingham and K. Smallbone, *Sci. Total Environ.*, 2000, **253**, 151–167.
[11] J. Clench-Aas, A. Bartonova, R. Klaeboe and M. Kolbenstvedt, *Atmos. Environ.*, 2000, **34**, 4737–4744.
[12] Y. Chan, R. Simpson, G. Mctainsh, P. Vowles, D. Cohen and G. Bailey, *Atmos. Environ.*, 1999, **33**, 3251–3268.
[13] J. Stedman, E. Linehan and B. Conlan, *Atmos. Environ.*, 2001, **35**, 297–304
[14] J. Stedman, *Atmos. Environ.*, 2002, **36**, 4089–4101.
[15] R. Derwent, P. Simmonds and W. Collins, *Atmos. Environ.*, 1994, **28**, 2623–2637.

Table 4 Contribution of PM_{10} related to road traffic at 17 UK cities in 1996 using CO as a tracer of traffic related emissions[14]

City	% Contribution of road traffic to PM_{10}		
	Winter	Summer	Annual
Wolverhampton	67.9	36.2	52.1
Leeds	58.8	40.0	49.4
Nottingham	54.1	22.6	38.3
Newcastle	46.5	22.4	34.4
Manchester	43.6	35.5	39.5
Edinburgh	41.4	20.5	30.9
Bristol	37.0	29.6	33.3
London Kensington	36.9	28.2	32.6
Hull	36.8	7.8	22.3
Birmingham Centre	36.7	26.9	31.8
Liverpool	34.2	46.4	40.3
Cardiff	33.7	23.2	28.5
Birmingham East	30.8	20.4	25.6
London Bexley	30.8	33.1	32.0
Southampton	30.0	51.6	40.8
London Brent	22.7	19.0	20.9
Swansea	14.7	20.0	17.4

the PM_{10} concentration corresponding to the appropriate seasonal background of CO, and hence to a situation with no influences of motor traffic. The concentration of PM_{10} thus arrived at corresponded to the PM_{10} in air which does not arise from road traffic. Subtraction of this concentration from the measured average concentration for the site and season provided the study with an estimate of the concentration of PM_{10} generated from road traffic. Table 4 shows the percentage contribution of the measured PM_{10} related to road traffic sources for the 17 urban background sites for winter, summer, and the full year based upon the CO receptor modelling technique.

Stedman *et al.*[13] further developed the receptor model for predicting future concentrations of PM_{10} for a central London site and later for a central Belfast and Bury roadside site.[14] The early part of 1996 provided an unusual period of air pollution with an exceptional prevalence of high particle concentrations resulting mainly from transport of air pollutants from the European continent.[16] This presented an opportunity to develop a method to discriminate the secondary contribution to airborne particulate matter from particles arising from other sources.

The receptor modelling technique developed divided the measured daily mean PM_{10} concentration at a monitoring site into three components:

- Primary combustion particles;
- secondary particles;
- other particles—assumed to consist largely of coarse particles (diameters 2.5–10 μm), which are primarily made up of re-suspended dust and sea salt.

[16] J. Stedman, *The Secondary Particle Contribution to Elevated PM_{10} Concentrations in the UK*, Report 20008001/007, AEAT–2959, AEAT, NETC, 1997.

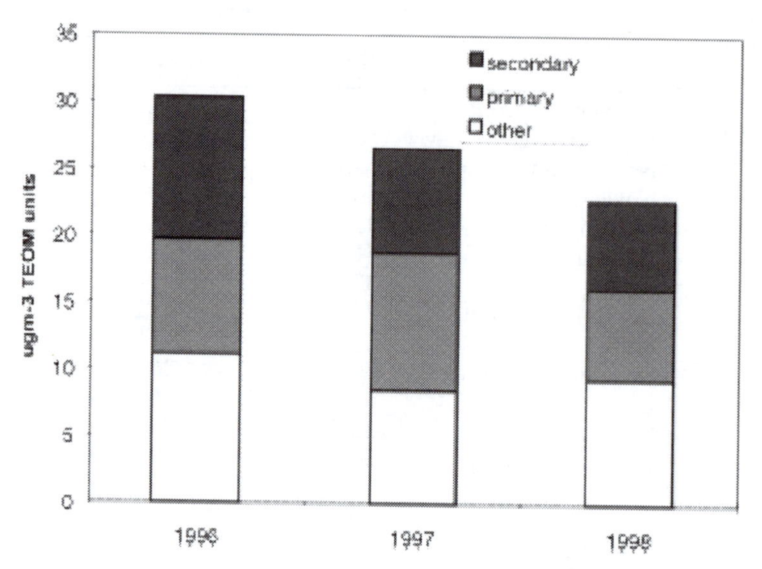

Figure 1 Estimated contributions to annual mean PM_{10} at London Bloomsbury in 1996–1998[10]

NO_x measurements were used as an indicator for primary combustion particles and rural sulfate measurements were used as an indicator for secondary particles. A multiple regression analysis was carried out to determine the coefficients *A* and *B* for primary combustion, and secondary particle concentrations, such that:

$$[\text{measured PM}_{10}\ (\mu\text{g m}^{-3})] = A\ [\text{measured NO}x\ (\mu\text{g m}^{-3})] + B\ [\text{measured sulfate}\ (\mu\text{g m}^{-3})] + C\ (\mu\text{g m}^{-3}) \quad (1)$$

The daily mean concentration of the other particles was determined by difference. It was expected that the concentrations of primary combustion and re-suspended dust particles be correlated to some extent since both are related to road traffic activity. A weak correlation between the daily increment of roadside concentrations of coarse particles and $PM_{2.5}$, above that recorded at a nearby urban background monitoring site, had been previously demonstrated by APEG for a roadside monitoring site in central London.

A key uncertainty within the receptor modelling method is the assignment of the residual PM_{10}, remaining after the assignment of primary combustion and secondary particle contributions, to the other particle fraction. An examination of the difference between measured PM_{10} and $PM_{2.5}$ concentrations enabled the researchers to confirm the assignment of the bulk of this residual to coarse particles with diameters in the range 2.5 to 10 μm. The estimated contributions to annual mean PM_{10} for the three assigned PM fractions at the central London site in 1996–1998 are provided in Figure 1.

High-resolution spatial maps of concentrations of NO_2 at locations in the UK have been estimated by Stedman *et al.* by using empirical modelling.[17,18] The

[17] J. Stedman, K. Vincent, G. Campbell, J. Goodwin and C. Downing, *Atmos. Environ.*, 1997, **31**, 3591–3602.

[18] J. Stedman, J. Goodwin, K. King, T. Murrells and T. Bush, *Atmos Environ.*, 2001, **35**, 1451–1463.

Table 5 Percentage contribution to the NO_X and NO_2 concentrations at urban and suburban sites

	Interpolated rural	*Urban and suburban*	*NO_X emissions from major roads*
NO_X	17.4	59.3	23.3
NO_2	18.1	63.3	8.6

relationships incorporated the data from air monitoring sites within the UK air monitoring network. These included urban background, urban centre, suburban or rural, and roadside or kerbside sites. Each category of monitoring site was considered separately. The impacts of major NO_2 sources upon the rural sites were considered minimal where such sources were at a distance of more than 10 km. It was also assumed that the road traffic emissions of NO_X determined the roadside increment of NO_X . Thc NO_2 concentrations at the sites were then estimated, based upon a linear relationship with the measured NO_X concentration that depended upon the category of monitoring site. Table 5 shows the percentage contribution from three components making up NO_X and NO_2 concentrations estimated by Stedman *et al.*[17]

In France, the ExTra index,[19] produced by the French Scientific Center for Building Physics and the French National Institute for Transport and Safety Research (INRETS), has been employed to estimate ambient concentrations of transport-related pollutants in front of the work and living places of urban dwellers.[20] Calculation of the index is based upon two dispersion models and evaluation of the index upon the comparison with NO_X concentrations measured in various street canyons with those calculated with the index. The two series of calculated and measured concentrations show a high correlation and good agreement for the majority of sites (Grenoble, r = 0:90, R = 0:75; Nice, r = 0:95; R = 0:91; Paris, r = 0:89; R = 0:89 and Toulouse, r = 0:89; R =0:86). The ExTra index was thus considered suitable to accurately classify sites according to their NOx exposure and concluded to reduce misclassification of transport-related air pollutants.

Estimates of Concentrations of Air Pollutants Attributable to other Transport Sources

In a report[21] prepared for DETR by AEA Technology the impact on air quality of postulated growth projections of UK regional airports was assessed. Although the study objective was to assess future growth projections of the airports, the complexities associated with assigning estimated pollutant concentrations to a specific source type are examined in the process. Twenty three regional airports were judged for the pollutants NO_2 and PM_{10}. Neither of these pollutants is unique to airports, being emitted by various sources, in particular road vehicles and power stations. On an airport, the chief emitters of these pollutants are aircraft, road vehicles and airside support vehicles/plant.

[19] C. Sacre, M. Chiron and J.-P. Flori, *Sci. Total Environ.*, 1995,**169**, 63–69
[20] P. Reungoat, M. Chiron, S. Gauvin, Y. Moullec and I. Momas, *Environ. Res.*, 2003, **93**, 67–78.
[21] B. Underwood, S. Brightwell, M. Peirce and C. Walker, *Air Quality at UK Regional Airports in 2005 and 2010*, Report AEAT/ENV/R/0453, Issue 2, 2001.

Two classes of receptors were considered for the model:

1. Background' receptors — these are far enough from roads that they do not receive a significant above-background contribution from the nearby road links.
2. 'Near-road' receptors — these are close enough to the nearest road links that they receive a significant above-background contribution from those links.

For the first class, 'background' receptors, the concentration of the specified pollutant was considered as having two components:

- The contribution from 'aircraft-related' sources, defined as aircraft exhaust emissions, APU (Auxiliary Power Unit) emissions and emissions from airside support vehicles.
- The contribution from all other sources, including landside road vehicles and sources that contribute to general background levels in the region of the airport.

For the second class, 'near-road' receptors, the concentration of the specified pollutant was considered as having these two components in addition to the contribution from the nearby road link(s), which was considered separately.

For the contribution from aircraft-related sources, the concentration estimate is based on an evaluation of the emissions and the calculation of concentrations arising from those emissions using conventional atmospheric dispersion modelling, The contribution to annual-mean concentrations from all other sources was estimated using the spatial distribution maps developed by Goodwin *et al.*[7] For NO_x, this represents the concentration at any point in the UK as arising from two components, the first related to local emissions (*i.e.* emissions in a 5×5 km square surrounding the point) and the second a background contribution taking into account more distant sources, obtained by interpolating rural monitoring data. The coefficient relating concentration to local emissions was empirically derived from analysis of national monitoring data. A similar procedure was used for PM_{10}.

Further investigations into the air quality impact of five options involving proposed changes to the infrastructure layout with an assigned level of activity at Edinburgh and Glasgow airports were also examined.[22] It is conventional to include aircraft emissions up to 1000 m in airport emission inventories, as it is argued that the emissions at greater height have a relatively smaller impact on ground-level airborne concentrations around the airport. In particular, emissions from climb-out and a substantial fraction of the emissions from initial climb and approach have an insignificant impact on local air quality. The dispersion modelling accounted for the varying impact of emissions as a function of height. For NO_x, the emissions from the take-off roll were estimated as a major contributor to the near-ground emissions. It was assumed that all aircraft took off at 100% thrust, whereas in practice most large jets often take off at reduced thrust. Thus the take-off emissions are more likely to have been overestimated than underestimated. For PM_{10}, the large contribution from aircraft brake and tyre wear was also apparent.

[22] B. Underwood, M. Peirce and C. Walker, *Regional Air Services — Part 3 Study: Air Quality Modelling for Scotland*, Report AEAT/ENV/0896, 2002.

Table 6 Estimates for the mass emissions and the change in emissions (both mass in kg day^{-1} and percentage) for two ferry technologies compared with on-land commute trips under two average ferry capacity conditions in the San Francisco Bay Area for the Larkspur Route[31]

	Existing ferries					*Ferries with alternative technology*				
		Change from on-land 25% occupancy		*Change from on-land 75% occupancy*			*Change from on-land 25% occupancy*		*Change from on-land 75% occupancy*	
Pollutant	*Emissions (kg day^{-1})*	*(kg day^{-1})*	*(%)*	*(kg day^{-1})*	*(%)*	*Emissions (kg day)$^{-1}$*	*(kg day)$^{-1}$*	*(%)*	*(kg day)$^{-1}$*	*(%)*
NO_x	430	420	5300	400	−1700	41	33.07	41.7	17	72
NMHC	8.9	1.4	19	−13.5	−60	0.47	−7.0	−94	−22	−98
CO	82	0.79	1	−160	−68	6.2	−75	−92	−240	−97
PM	9.9	2.7	37	−12	−54	0.46	−6.9	57	−21	−98

Much of the emissions from marine transportation may occur away from heavily inhabited areas and for this reason, and reasons related to analytical difficulties in such areas, sparse data are available for source apportionment of these source types. Nevertheless, marine engines are a significant and growing source category in some locations and there are concerns over the marine transport sulfur dioxide emissions.[23–27] In a recent report[28] the emissions of polycyclic aromatic hydrocarbons (PAH) were estimated as ten times higher per unit of energy from the burning of heavy oil (marine residual fuel) than from marine gas oil. In addition, the PAH emissions from a ship using marine residual fuel were considered about 30 times higher per unit of energy than those from a heavy diesel-driven vehicle. This means that if the output of a ship's engine is, say, 40 times that of an automotive diesel vehicle, the emissions from a fairly large vessel entering port will correspond to those from about 1200 heavy diesels. In European waters the emissions from international shipping have been estimated as 2.6 million tonnes of SO_2, and 3.6 million tonnes of NO_x (as NO_2) in 2000,[27] and concerns over these levels have led the European parliament to vote to strengthen measures to reduce the emissions from marine fuels. These include a reduction of the sulfur content of the fuels from a maximum permitted concentration of 1.5 to 0.5%. However, the vote has come under criticism since it demanded no haste with the target year for these measures set at 2013.[29]

Comparisons of emissions from marine and on-land passenger travel have been investigated. Using data gathered in the San Francisco Bay Area, Corbett and Farrell[30] (2002) and Farrell *et al.*[31] (2003) have modelled emissions from three passenger ferries and the matching on-land travel that would be used by commuters if the ferry service were not available. The study evaluated the replacement of existing uncontrolled diesel engines installed onboard three specific ferries with seven alternative technologies that could be used to control emissions. Emissions were estimated for the existing technology and each alternative based on current ferry service (*i.e.* number, length, and speed profiles of existing commute trips). In addition to this 'waterside' analysis, the study also explored the emissions that would occur if current ferry passengers commuted by a land route instead (*i.e.*, the 'landside' analysis). Activity based methods were employed to estimate the contributions of the marine and road sources to the concentrations of NO_X, NMHCs, PM, CO, SO_2, and CO_2. Table 6 presents the estimates for the mass emissions and the change in emissions (both mass and percentage) for two ferry technologies compared with on-land commute trips under two average ferry capacity conditions.

[23] J. Corbett and P. Fischbeck, *Science*, 1997, **278**, 823–824.
[24] M. Lawrence and P. Crutzen, *Nature*, 1999, **402**, 167–170.
[25] K. Capaldo, J. Corbett, P. Kahsibhatla, P. Fischbeck and S. Pandis, *Nature*, 1999, **400**, 743–746.
[26] J. Corbett and P. Fischbeck, *Environ. Sci, Technol.*, 2000, **34**, 3254–3260.
[27] ENTEC, *Quantification of Emissions from Ships Associated with Ship Movements between Ports in the European Community*, www.europa.eu.int/comm/environment/air/background.htm-transport, 2002.
[28] J. Ahlbom and U. Duus, *Rent Skepp Kommer Lastat*, www. Gronkemi.nu/skepp.html, 2003.
[29] C. Agren, *Acid News*, 2003, **3**, 1.
[30] J. Corbett and A. Farrell, *Trans. Res.*, 2002, **D7**, 197–211
[31] A. Farrell, D. Redman J. Corbett and J. Winebrake, *Trans. Res.*, 2003, **D8**, 343–360.

3 Health Effects of Transport-related Air Pollutants

Only a rather brief overview can be given here of the health effects of air pollutants to serve as an introduction to the subsequent sections on quantification of effects. More extensive and authoritative reviews[32,33] are available.

Carbon Monoxide

This gas exerts its toxic effect in humans through binding with haemoglobin in the blood. It binds reversibly but more strongly than oxygen, thereby reducing the oxygen carrying capacity of the blood. Normally, non-smoking adults have carboxyhaemoglobin levels (*i.e.* the percentage of haemoglobin in the form of carboxyhaemoglobin) of 1% or less whilst cigarette smokers may have levels ranging from 4 to 15% depending on the number of cigarettes smoked. Increases in the carboxyhaemoglobin level place stress on the heart and patients with angina experience episodes of chest pain earlier during exertion when their carboxyhaemoglobin level is elevated. To prevent this effect, air quality standards are set at a level that should ensure that carboxyhaemoglobin levels remain below 2.5%, which is below the threshold at which exacerbation of angina symptoms is observed. Since the advent of catalytic converters on petrol cars, air quality standards for carbon monoxide are rarely exceeded in developed countries, although exceedences are still very likely in the less developed world. Time series epidemiological studies in which day-to-day changes in hospital admissions and mortality are related to day-to-day changes in air pollutant concentrations show effects on both mortality and hospital admissions for cardiovascular disease when carbon monoxide concentrations are elevated. However, the extent to which this observation is causally linked with carbon monoxide exposure as opposed to being due to the effects of other traffic-generated pollutants is very hard to discern.

Sulfur Dioxide

Historically, elevated concentrations of sulfur dioxide were always associated with high concentrations of particulate matter since both arose simultaneously from the combustion of coal. It was, for many years, believed that there was an interaction, most probably synergistic in nature between smoke (combustion generated particles) and sulfur dioxide in eliciting adverse effects on health. More recent research has tended to indicate that the two act essentially independently and, therefore, effects of sulfur dioxide should be considered in their own right.

The most marked effect of sulfur dioxide on health, which forms the basis of the UK air quality standard for sulfur dioxide, is the induction of broncho-constriction (*i.e.* narrowing of the upper airways) in asthmatics when exposed to concentrations in excess of around 200 ppb for a few minutes. Such results,

[32] S. T. Holgate, J. M. Samet, J. S. Koren and R. L. Maynard (eds.), *Air Pollution and Health*, Academic Press, 1999.

[33] R.E. Hester and R.M. Harrison (eds.), *Air Pollution and Health*, Issues in Environmental Science and Technology, No. 10, Royal Society of Chemistry, Cambridge, 1999.

arising from controlled clinical human challenge studies, give little indication of the public health impact, which has been studied through various types of epidemiological research, involving detailed studies of the symptoms of individuals in relation to day-to-day changes in sulfur dioxide, as well as time series studies involving mortality and hospital admissions statistics for entire urban populations. The latter demonstrate clear effects upon hospital admissions for respiratory disease (most probably mainly in those with pre-existing disease such as asthma or chronic obstructive pulmonary disease) as well as increases in mortality due to both respiratory and cardiovascular disease.

Nitrogen Dioxide

Despite rather intensive research in recent years, adverse effects of nitrogen dioxide on public health remain rather controversial, with some leading workers taking the view that adverse effects have not been clearly demonstrated and that apparent connections between nitrogen dioxide and adverse effects in epidemiological studies are probably the result of co-exposure to other pollutants, such as particulate matter, which tend to vary in concentration in the atmosphere in a similar manner to nitrogen dioxide.

Clinical human exposure studies have shown reductions in lung function in asthmatics exposed to nitrogen dioxide and these form the basis of air quality standards for short-term (acute) exposures. There is also evidence that nitrogen dioxide may enhance the response to airborne allergens such as grass pollen. The results of epidemiological studies are far less consistent in relation to acute effects, and whilst some show effects on both hospital admissions and mortality, other studies have been unable to demonstrate such effects.

The World Health Organisation recommends an air quality guideline also in respect of chronic (long-term) exposures to nitrogen dioxide. This is based on epidemiological studies that show an increased rise of respiratory illness of children at fortnightly average concentrations above about 30 $\mu g\ m^{-3}$ measured as a personal exposure, which includes exposures due to indoor sources such as gas cooking, which can cause a major elevation in nitrogen dioxide concentrations.

Particulate Matter

As outlined above, current thinking is that the adverse effects of particulate matter are essentially independent of those of sulfur dioxide, with which it is now far less frequently associated in the atmosphere than in times when domestic coal burning was a major source of pollution in developed countries.

Evidence for the adverse effects of particulate matter comes mainly from population-based epidemiological studies and is the clearest and strongest for any of the classical air pollutants. Whilst the mechanisms are still a subject of extensive research, the effects are well recognized and quantified. Short-term effects (*i.e.* of daily exposures) have been elucidated primarily through time series epidemiological studies that have shown very clear associations between exposure to airborne particles measured by a variety of metrics, including PM_{10} and $PM_{2.5}$, and adverse effects on health, most specifically mortality and hospital admissions for both respiratory and cardiovascular disease. Studies have also extended to less severe health outcomes (which are important because they affect larger numbers of people), such as exacerbated symptoms and increased bronchodilator use by asthmatics, and reduced activity days. Whilst such effects are demonstrable in epidemiological studies, they are far less easy to quantify in terms of the impact on public health. Whilst laboratory-based toxicological studies appear to indicate important influences of both particle size and chemical composition upon toxicity, these studies are carried out at exposures far in excess of environmental levels, and the community-based epidemiological studies do not appear to show a strong influence of location and hence size distribution and particle chemistry on the magnitude of effects per unit concentration of particles.[34] There is, however, some evidence from the APHEA II study and work by Harvard University[35] that road traffic-generated particles may be more potent than particles from other sources in eliciting adverse effects on health.

Long-term studies in which the effects of living in cities with different levels of air pollution by particulate matter are linked to mortality rates indicate a very important effect of chronic exposure to airborne particles upon life expectancy. The results of such studies have been used extensively in estimating the public health impacts of particulate matter exposure and will be explored later.

Ozone

Ground-level ozone is a pollutant that is not emitted in significant amounts by any source of air pollution. It is, however, a secondary pollutant formed within the atmosphere from chemical reactions of oxides of nitrogen and volatile organic compounds in sunlight. Transport-generated air pollutants therefore play a major role in its formation.

Ozone is a highly reactive oxidant gas that appears to affect both asthmatics and non-asthmatic subjects equally. It has been clearly shown to cause a reduction in lung function for exposure to elevated levels over periods of six to eight hours. Epidemiological studies have sought to elucidate the effects of ozone on public health; these have involved both panel studies of a small number of individuals who have shown increased respiratory systems with increasing concentrations of ozone, and larger city-based time series studies, which show increases in both morbidity and mortality with increasing ozone concentrations. What is less clear, however, is whether there is a threshold concentration below which effects do not occur within the population. This is difficult to discern from the epidemiology, particularly because of the complex mixture of pollutants to which individuals are exposed in urban areas, but the prevailing current view is that the existence of a threshold is unlikely.

Chemical Carcinogens

Whilst cancer is high in the public mind as an adverse outcome of air pollutant exposure, the evidence of an effect from population-based studies is very weak.

[34] R. M. Harrison and J. Yin, *Sci. Total Environ.*, 2000, **249**, 85.
[35] F. Laden, L. M. Neas, D. W. Dockery and J. Schwarz, *Environ. Health Perspect.*, 2000, **108**, 941.

However, extrapolation of data collected with occupationally exposed populations indicates a small but significant cancer risk from exposure to urban concentrations.

Probably the most important air pollutants in this regard are the polycyclic aromatic hydrocarbons, which are formed in all combustion processes. The best known of these compounds, benzo[*a*]pyrene is typically used as a marker for the complex mixture of substances and is itself responsible for a substantial proportion of the carcinogenicity of the entire mixture.[36] Exposure to airborne PAH leads to an increase in lung cancer incidence, and the World Health Organisation and USEPA publish unit risk factors from which the increased risk of cancer can be calculated from exposure concentrations.

There are a number of other chemical carcinogens associated with emissions from transport sources for which air quality standards have been set. Most important amongst these are the gaseous pollutants benzene and 1,3-butadiene, which have been associated in occupational exposures with an increased risk of leukaemia and lymphoma, respectively. Several trace metals, including nickel, chromium in its Cr(VI) oxidation state and arsenic, also have carcinogenic properties but these are generally considered to be only a minor cause of cancer risk to the general public in developed countries. Emissions of nickel from ships burning heavy fuel oil present a poorly quantified source of risk.

4 Methods of Quantification of the Effects of Air Pollution on Health

To conduct cost–benefit analyses for air pollution abatement and to compare the impacts of different pollutants and sources upon health, it is often considered desirable to quantify the impacts of a specific pollutant or source upon public health. The sources of information from which to conduct such an exercise are severely limited. Whilst the clinical controlled challenge studies provide excellent information on thresholds (amongst those subjects who are studied, which may not include the most susceptible) and mechanisms, they give no useful information on the effects on the population as a whole. For this it is necessary to use data from epidemiological studies of kinds that study large numbers of individuals either through using population-based statistics (*e.g.* mortality rates within a city) or by enrolling large numbers of individual subjects (*e.g.* the American Cancer Society Study[37] which enrolled more than half a million subjects) into a cohort study.

Acute (Short-term) Effects

The time series epidemiological studies that link day-to-day changes in air pollutant concentrations with changes in daily symptoms, hospital admissions or mortality provide the strongest basis from which to estimate acute effects at the population level. Such studies provide coefficients typically in the form of a percentage increase in a health outcome per unit increase in pollutant concentration.

[36] Expert Panel on Air Quality Standards, *Polycyclic Aromatic Hydrocarbons*, Stationery Office, 1999.
[37] C. A. Pope III, R. T. Barnett, M. J. Thun, E. G. Calle, D. Krewski, K. Ito and G. D. Thurston, *J. Am. Med. Assoc.*, 2002, **287**, 1132.

Table 7 Numbers of deaths and hospital admissions for respiratory diseases affected per year by PM_{10}[a] and sulfur dioxide in urban areas of Great Britain[38]

Pollutant	*Health Outcomes*	*GB urban*
PM_{10}	Deaths brought forward (all cause)	8100
	Hospital admissions (respiratory) brought forward and additional	10500
SO_2	Deaths brought forward (all cause)	3500
	Hospital admissions (respiratory) brought forward and additional	3500

[a]PM_{10}: particulate matter generally less than 10 μm in diameter. Estimated total deaths occurring in urban areas of GB per year = c. 430 000. Estimated total admissions to hospital for respiratory diseases occurring in urban areas of GB per year = c. 530 000.

Table 8 Numbers of deaths and hospital admissions for respiratory diseases affected per year by ozone in both urban and rural areas of Great Britain during summer only[38]

Pollutant	*Health outcomes*	*GB, threshold = 50 ppb*	*GB, threshold = 0 ppb*
Ozone	Deaths brought forward: all causes	700	12 500
	Hospital admissions (respiratory) brought forward and additional	500	9900

Thus, for example, most studies of the impact of airborne particulate matter on mortality have generated exposure–response coefficients of approximately 0.5-1% increase in mortality per 10 μg m^{-3} of PM_{10}. At a simplistic level, the data may be used as in the following example: in a city of one million people with an average PM_{10} concentration of 20 μg m^{-3} and an annual mortality rate of 1000 per 10^5 population, how many deaths brought forward are attributable annually to particulate matter exposure if the exposure–response coefficient is 0.5% per 10 μg m^{-3}? Since the exposure–response relationship for PM_{10} within concentration ranges currently experienced appears to be linear without a no-effect threshold, rather than calculating daily numbers of deaths based upon the 24-hour pollutant concentrations used in the time series epidemiological studies, it is permissible to work with annual mean concentrations, which greatly simplifies the calculation. Therefore, for a coefficient of 0.5% per 10 μg m^{-3} with an annual mean concentration of 20 μg m^{-3} an increase in the annual mortality rate of 1% is implied. Since the annual mortality rate in the city is 1000 per 10^5 population, this implies a total annual number of deaths of 10 000 in the population of one million. One percent of this is 100 deaths brought forward due to particulate matter exposure.

The next stage of sophistication in such calculations involves breaking down the geographic area of a city or country into grid squares on which such calculations are conducted individually and then summed to give a total impact across the domain of the study. This then allows the use of spatially varying air pollutant concentrations as well as different death rates for areas with differing age structures and life expectancy.

One of the pioneers of this approach was the UK Department of Health Committee on the Medical Effects of Air Pollutants, which used such a gridded approach to air pollutant concentrations (but not to mortality and hospital admission rate data) to calculate numbers of deaths and hospital admissions affected by air pollutants per year.[38] The Committee took a cautious approach, applying its calculations only to urban areas of Great Britain since the exposure–response coefficients had been determined on urban populations and therefore any extrapolation to rural populations involved a significant uncertainty. Additionally, since there are significant differences between the air pollution climate of Northern Ireland and of Great Britain, and the exposure–response coefficients were unavailable for Northern Irish cities, the calculations were restricted to England, Wales and Scotland. The results of the study for PM_{10} and sulfur dioxide appear in Table 7. Additionally, Table 8 shows the results of calculations for ozone, which were made both with a threshold concentration of 50 ppb and for a zero threshold. For ozone, the calculations included both urban and rural areas for the summer months only and were based on an exposure–response coefficient for eight hourly mean ozone concentrations, which required a more sophisticated means of calculation than those involving the other pollutants, such that individual daily calculations where conducted. COMEAP also calculated effects of nitrogen dioxide upon hospital admissions as a sensitivity study but cautioned that this probably involved double-counting of the effects of the pollutant mixture.[38] Consideration was given as to whether there were appropriate coefficients to calculate the effects of other pollutants such as carbon monoxide but the conclusion was drawn that the information base was inadequate to conduct appropriate calculations.

Since the publication of the COMEAP report, the general approach used has been widely applied in calculating public health impacts of point sources of air pollutants. Such an approach has been used in estimating the externalities of power generation and waste management options for example. In the near field (*i.e.* within a few kilometres of the source) it involves using dispersion models to calculate ground-level concentrations on a gridded basis, which may be then used for calculating public health impacts in a manner directly analogous to that used for calculating national impacts as in the COMEAP report. The uncertainties in such calculations are, however, considerably higher than those for whole urban populations and COMEAP has issued a cautionary statement indicating the caveats that apply to calculations conducted in this manner.[39]

Whilst concentrations of the primary (emitted) pollutants from point sources fall off rapidly with distance and, therefore, only near-field calculations are warranted, secondary pollutants such as particulate nitrate and sulfate and gas-phase ozone form as a result of the emissions on far greater distance scales.[40] Whilst there are very considerable uncertainties attached to any such calculations, some workers have used numerical model studies to estimate the impact of the emissions of a specific point source upon concentrations of secondary pollutants

[38] Department of Health, Committee on the Medical Effects of Air Pollutants, *Quantification of the Effects of Air Pollution on Health in the United Kingdom*, 1998.

[39] Department of Health, Committee on the Medical Effects of Air Pollutants, http://www.doh.gov.uk/comeap/statementsreports/areaemissions.htm

[40] R. M. Harrison, in *Pollution, Causes, Effects and Control*, 4th Edn., R. M. Harrison (ed.), Royal Society of Chemistry, Cambridge, 2001.

Table 9 Number of cases attributed to traffic pollution in three European Countries[44]

	Estimated attributable number of cases or days Traffic-related air pollution (PM_{10})		
Health outcomes	*Austria*	*France*	*Switzerland*
Long-term mortality (adults ⩾ 30 years)	2400	17 600	1800
Respiratory hospital admissions (all ages)	1500	7700	700
Cardiovascular hospital admissions (all ages)	2900	11 000	1600
Chronic-bronchitis incidence (adults > 25 years)	2700	20 400	2300
Bronchitis (children < 15 years)	21 000	250 000	24 000
Restricted activity days in adults ⩾ 20 years (in millions)	1.3	13.7	1.5
Asthmatics: asthma attacks (children < 15 years)	15 000	135 000	13 000
Asthmatics: asthma attacks (adults ⩾ 15 years, person days)	40 000	321 000	33 000

at large distances from the point of emissions.[41] Theoretically, if such concentrations are calculable it is then possible to use the same gridded approach to calculate public health impacts. Such an approach involves substantial uncertainties in the calculations of secondary pollutant concentrations as well as the uncertainties inherent in applying the exposure–response coefficients. Overall, therefore, impacts calculated in this way are open to far greater uncertainty limits than the more restricted calculations conducted by COMEAP.

Chronic Effect

Cohort epidemiological studies conducted in North America have demonstrated a relationship between long-term average urban concentrations of airborne particulate matter and death rates, when corrected for other risk factors. The most influential of these studies have been the Harvard Six Cities Study[42] and the American Cancer Society Study.[37] Both of these studies recruited large numbers of individual adult subjects who were followed forward in time over a period of up to 20 years, their death rates being recorded according to the city in which they resided and related to the air pollutant concentrations. Both studies

[41] A. Rabl and J. Spadaro, in *Environmental & Health Impacts of Solid Waste Management Activities*, R. E. Hester and R. M. Harrison (eds.) Issues in Environmental Science and Technology, No. 18, Royal Society of Chemistry, Cambridge, 2002.

[42] D. W. Dockery, C. A. Pope III, X. Xu, J. D. Spengler, H. J. Ware, M. E. Fay, B. G. Ferris, Jr. and F. E. Speizer, *New Engl. J. Med.*, 1993, **329**, 1753.

Table 10 California annual morbidity effects in diesel $PM_{2.5}$ health effects studies (adapted from Reference 45)

Health endpoint	*$PM_{2.5}$ concentration ($\mu g\ m^{-3}$)*	*Incidence (cases per year)*
Chronic bronchitis	1.8[a]	1791
(age 27+)	0.81[b]	801
		2592
Hospital admissions		
Chronic obstructive pulmonary disease	1.8[a]	441
(age 65+)	0.81[b]	198
		638
Pneumonia	1.8[a]	537
(age 65+)	0.81[b]	241
		778
Cardiovascular	1.8[a]	1297
(age 65+)	0.81[b]	583
		1881
Asthma	1.8[a]	209
(age 64−)	0.81[b]	94
		303
Asthma-related ER visits	1.8[a]	1155
(age 64−)	0.81[b]	518
		1673

[a] From Cal/EPA (2000). [b] Estimated conversion of diesel NO_X emissions into $PM_{2.5}$ nitrate using 1999 California air quality data. Likely to be an underestimate because of NH_4NO_3 loss.

showed a significant relationship between urban death rates and airborne particulate matter concentrations, which have been taken to indicate a causal relationship between long-term exposure to particulate matter and reduction in life expectancy. Such a relationship is not directly useful in calculating the public health impacts, and in the UK the Institute of Occupational Medicine has applied a life table approach to calculating reductions in life expectancy as a result of particulate matter exposure. The Committee on the Medical Effects of Air Pollutants which commissioned this study looked critically at the range of values, taking the view that the effects were most likely towards the lower end of the plausible range. For the population of England and Wales alive in 2000 it was thought most likely that a reduction of 1 $\mu g\ m^{-3}$ in $PM_{2.5}$ might lead to an increase of life expectancy of 1.5–3.5 days per person on average.[43] This is more than ten times greater than the average increase in life expectancy from abatement of particulate matter based on the short-term effects upon mortality (as indicated by the time series studies) and assuming that for each death brought forward the loss of life would be within the range of two to six months. For a birth cohort born in 2000 and followed up for their lifetime, their gain in life expectancy for an abatement of 1 $\mu g\ m^{-3}$ of $PM_{2.5}$ was estimated as between 0.5 and 4.5 weeks.[43]

Attribution of Effects Estimates to Road Traffic Sources

Kunzli and co-workers[44] have used coefficients from both time series and cohort

[43] Department of Health, Committee on the Medical Effects of Air Pollutants, http://www.doh.gov.uk/comeap/statementsreports/longtermeffects.pdf

Table 11 California annual mortality effects in diesel $PM_{2.5}$ health effects studies (adapted from Reference 45)

Health endpoint	*$PM_{2.5}$ concentration ($\mu g\ m^{-3}$)*	*Incidence (cases per year)*
Long-term exposures mortality		
Indirect diesel PM	1.8[a]	1985
Indirect PM	0.81[b]	895
		2880
Short-term exposure mortality		
Direct diesel PM	1.8[a]	459
Indirect diesel PM	0.81[b]	206
		665

Note: Long-term study should be used alone rather than considering the total effect to be the sum of estimated short- and long-term effects, because summing creates the possibility of double-counting a portion of PM-related mortality. [a] From Cal/EPA (2000). [b] Estimated conversion of diesel NO_X emissions into $PM_{2.5}$ nitrate using 1999 California air quality data. Likely to be an underestimate because of NH_4NO_3 loss.

epidemiological studies to estimate a public health impact of both all outdoor and traffic-related air pollution for European countries. Because of the frequent correlations observed between primary air pollutants, Kunzli and co-workers took the view that epidemiological studies cannot strictly allocate observed effects to single pollutants and therefore a pollutant by pollutant assessment would grossly overestimate the impact. They therefore used PM_{10} as a useful indicator of several sources of outdoor air pollution such as fossil fuel combustion. Their model, following the pattern outlined above, then used the exposure–response function, the frequency of the health outcome and the level of exposure to estimate an attributable number of cases. To drive the population exposure distribution, annual mean concentrations of PM_{10} were modelled at a spatial resolution of 1 km^2 (or 4 km^2 in France). The health impact of air pollution exposure below 7.5 $\mu g\ m^{-3}$ of $PM_{10}$was discounted, not on the basis of any assumed threshold but because no studies were available below this concentration, which corresponds very approximately to natural background PM_{10}. Using an emission–dispersion model for particulate matter, the proportion of PM_{10} exposure attributable to traffic emissions was estimated. This proportion ranged from 28% at an annual mean PM_{10} of 10–15 $\mu g\ m^{-3}$ up to 58% for PM_{10} greater than 40 $\mu g\ m^{-3}$. These percentages look relatively high in the context of data from the United Kingdom (see above). Results were presented for Austria, France and Switzerland and appear in Table 9 in relation to both total outdoor air pollution and traffic-related air pollution. Estimates of mortality effects were based on the long-term rather than short-term impacts and used coefficients from the Harvard Six Cities Study[42] and American Cancer Society Study.[37] The fact that these estimates are based purely on exposure to PM_{10} implies that they are most probably under-estimates, particularly as the effects of the secondary pollutant ozone, which is heavily influenced by road traffic emissions, was excluded from the analysis. Whilst not addressed explicitly in the paper, it appears that secondary particulate matter derived from road traffic emissions was most probably included within the estimates.

Lloyd and Cackette[45] have reviewed the environmental impact and control of automotive diesel engines. Having calculated concentration levels of directly emitted and secondary particulate matter arising from diesel engines for the year 2000, they use published exposure–response coefficients to calculate annual morbidity and mortality attributable to diesel $PM_{2.5}$ emissions in the state of California. The estimates appear in Tables 10 and 11.

5 Recent Investigations of the Impacts of Transport Emissions upon Human Health

There have been increasing numbers of investigations of the impacts of transport emissions upon health.[46–58] The majority of these investigations have concentrated upon the relationships of road transport emissions with respiratory illnesses in urban areas. Of particular interest are respiratory illnesses of children in relation to their exposure to traffic. Other investigations have considered cohorts of random population samples, or alternatively of relationships of hospital admissions with exposure to traffic. In some instances the exposure is quantified as traffic density and/or in others as exposure to air pollutants commonly generated by traffic emissions. Table 12 summarizes recent investigations.

Although most of these investigations add to the growing body of evidence incriminating traffic emissions in respiratory disease, the findings are by no means conclusive. In an editorial presented in the *European Respiratory Journal*, Brunekreef and Sunyer[58] consider the research of Nicolai *et al.*[49] and Lee *et al.*[50]

44 N. Kunzli, R. Kaiser, S. Medina, M. Studnicka, O. Chanel, O. Filliger, M. Herry, F. Horak, Jr., V. Puybonnieux-Texier, P. Quenel, J. Schneider, R. Seethaler, J.-C. Vergnaud and H. Sommer, *Lancet*, 2000, **356**, 795.

45 A. C. Lloyd and T. A. Cackett, *J. Air and Waste Manage. Assoc.*, 2001, **51**, 809.

46 A. Venn, S. Lewis, M. Cooper, R. Hubbard and J. Britton, *Am. J. Respir. Crit. Care Med.*, 2001, **164**, 2177–2180.

47 R. McConnell, K. Bergane, F. Gilliland, S. London, T. Islam, W. Gauderman, E. Avol, H. Margolis and J. Peters, *Lancet*, 2002, **359**, 386–391.

48 N. Janssen, B. Brunekreef, P. Van Vliet, F. Aarts, K. Meliefste, H. Harssema and P. Fischer, *Environ. Health Perspect.*, 2003, **111**, 1512–1518.

49 T. Nicolai, D. Carr, S. Weiland, H. Duhme, O. von Ehrenstein, C. Wagner and E. von Mutius, *Eur. Respir. J.*, 2003, **21**, 956–963.

50 Y. Lee, C. Shaw, H. Su, J. Lai, Y. Ko, S. Huang, F. Sung and Y. Guo, *Eur. Respir. J.*, 2003, **21**, 964–970.

51 M. Brauer, G. Hoek, P. Van Vliet, K. Meliefste, P. Fischer, A. Wijga, L. Koopman, H. Neijens, J. Gerritsen, Kerkhof *et al. Am. J. Respir. Crit. Care Med.*, 2002, **166**, 1092–1098.

52 C.-Y. Yang, S.-T. Yu and C.-C. Chang, *J. Toxicol. Environ. Health*, 2002, **65**, 747–755.

53 P. Wilkinson, P. Elliott, C. Grundy, G. Shaddick, B. Thakrar, P. Walls and S. Falconer, *Thorax*, 1999, **54**, 1070–1074.

54 B. Oftedal, P. Nafstad, P. Magnus, S. Bjørkly and A. Skrondal, *Eur. J. Epidem.*, 2003, **18**, 671–675.

55 G. Hoek, B. Brunekreef, S. Goldbohm, P. Fischer and P. van den Brandt, *Lancet*, 2002, **360**, 1203–1209.

56 D. Fusco, F. Forastiere, P. Michelozzi, T. Spadea, B. Ostro, M. Arcà and C. Perucci, *Eur. Respir. J.*, 2001, **17**, 1143–1150.

57 P. Penttinen, K. Timonen, P. Tiittanen, A. Mirme, J. Ruuskanen and J. Pekkanen, *Eur. Respir. J.*, 2001, **17**, 428–435.

in relation to earlier work on traffic-related air pollution and childhood respiratory illnesses. The editorial highlights the difficulties in performing a direct comparison between the findings of relevant studies because of the differences in the construction of the asthma variables that were used in the analyses. It is the definition of asthma employed by the studies that may influence the extent of the association of the disease with the traffic emissions.

This type of influence can only serve to contribute to the differences that are evident in the conclusions of the studies. In addition, there are confounding factors (*e.g.*, socio-economic factors, diet *etc.*) that make the statistical interpretations complicated. For example, several of the studies included in Table 12 seem to suggest that there is no evidence of an association between impaired respiratory health and traffic exposure.[47,52,53] In the research of McConnell *et al.*[47] high ozone concentrations were associated, although not significantly, with the risk of developing asthma in children playing three or more sports. Furthermore, measured pollutant concentrations in these areas were inversely related to the rate of incident asthma which some might interpret as indicating that air pollution did not cause asthma. With regard to the study by Venn *et al.*,[46] an earlier study in Nottingham[59] had not shown an association between proximity to roads and wheezing.

Interestingly, the study by Venn *et al.*[46] showed a trend indicating an increase in wheezing with proximity to a main road for girls. However, even living very close to the road appeared to have no effect upon the reported wheezing of boys. In an earlier investigation by van Vliet *et al.*[60] a more pronounced relationship of truck traffic intensity and black smoke measured at the children's schools with chronic respiratory symptoms was also reported for girls than for boys. It was concluded that long-term exposure to traffic-related air pollution, in particular diesel exhaust particles, may increase chronic respiratory symptoms, especially in girls. Exposure suspicion bias could not be entirely ruled out by the investigators as an alternative explanation of the findings, although it does not seem likely that such bias would operate only in girls.[61] Further investigations are required to determine if, and then why, girls are more susceptible to traffic-derived pollution than boys.

There is also an argument for further investigations of particle number concentrations relevant to the health outcomes of exposure to traffic emissions.[57] To test the hypothesis that the high numbers of ultrafine particles in ambient air might provoke alveolar inflammation and subsequently cause exacerbations in pre-existing cardiopulmonary diseases, adult asthmatics were followed with daily peak expiratory flow (PEF) measurements and symptom and medication diaries for six months, while simultaneously monitoring particulate pollution in ambient air. The daily mean number concentration of particles, but not particle mass (PM_{10}, $PM_{2.5}$, or PM_1), was negatively associated with daily PEF deviations. The strongest effects were seen for particles in the ultrafine range.

[58] B. Brunekreef and J. Sunyer, *Eur. Respir. J.*, 2003, **21**, 913–915.

[59] A. Venn, S. Lewis, M. Cooper, R. Hubbard, I. Hill, R. Boddy, M. Bell and J. Britton, *Occup. Environ. Med.*, 2000, **57**, 152–158.

[60] P. van Vliet, M. Knape, J. de Hartog, N. Jannsen, H. Harssema and B. Brunekreef, *Environ. Res.*, 1997, **74**, 122–132.

Table 12 Recent investigations of the relationships of traffic exposure with respiratory illnesses

Ref.	*Population investigated*	*Surrogate for traffic exposure*	*Health determinant*	*Findings*
46	Case-control sample of 6147 primary schoolchildren (age 4 to 11 yr) and a random cross-sectional sample of 3709 secondary schoolchildren (age 11 to 16 yr) in Nottingham	Proximity of family home to the nearest main road	Parental questionnaires for primary schoolchildren (age 4 to 11 yr) (14), and self-completion questionnaires in secondary schoolchildren (age 11 to 16 yr)	Children living within 150 m of a main road, the risk of wheeze increased with increasing proximity by an odds ratio (OR) of 1.08 (95% CI 1.00–1.16) per 30 m in primary schoolchildren, and 1.16 (1.02–1.32) in secondary schoolchildren. Most of the increased risk was localized to within 90 m of the roadside. Among primary schoolchildren, effects were stronger in girls than boys ($p = 0.02$)
47	3535 children with no history of asthma were recruited from schools in 12 communities and were followed up for up to 5 years. 265 children reported a new diagnosis of asthma during follow-up.	• Air pollutant concentrations of ozone, PM_{10}, $PM_{2.5}$, NO_2 and acid vapour • Number of outdoor sports played	Newly physician-diagnosed asthma	In communities with high ozone concentrations, the relative risk of developing asthma in children playing three or more sports was 3.3 (95% CI 1.9–5.8), compared with children playing no sports. Sports had no effect in areas of low ozone concentration (0.8, 0.4–1.6). Time spent outside was associated with a higher incidence of asthma in areas of high ozone (1.4, 1.0–2.1), but not in areas of low ozone. Exposure to pollutants other than ozone did not alter the effect of team sports

48	Children attending 24 schools within 400 m of busy motorways	• Air pollution measurements • Proximity to road • Traffic counts	• Parent completed questionnaires • Sensitization to common allergens • Bronchial hyperresponsiveness (BHR)	Respiratory symptoms increased near motorways with high truck but not high car counts, and related to air pollutants. Lung function and BHR not related to pollution. Sensitization to pollen increased in relation to truck not car counts
49	Random sample of 7509 schoolchildren	• Traffic counts • Estimated concentrations of soot, NO_2 and benzene	• Skin prick tests • IgE measurements • Lung function with spirometry • Parent completed questionnaires	Traffic counts were associated with current asthma, wheeze and cough. In children with tobacco-smoke exposure, traffic volume was additionally associated with a positive skin-prick test. Cough was associated with soot, benzene and NO_2, current asthma with soot and benzene, and current wheeze with benzene and NO_2. No pollutant was associated with allergic sensitization. High vehicle traffic was associated with asthma, cough and wheeze, and in children additionally exposed to environmental tobacco smoke, with allergic sensitization
50	331,686 nonsmoking children that attended middle schools located within 2 km of 55 air-monitoring stations	• Measured concentrations of NO_X, SO_2, ozone, CO and PM_{10} from 66 monitoring stations	• Parent completed questionnaires	Physician-diagnosed allergic rhinitis was found to be associated with higher nonsummer (September–May) warmth and traffic-related air pollutants, including CO, NO_X and O_3. Questionnaire-determined allergic rhinitis correlated only with traffic-related air pollutants

Table 12 *(cont.)*

Ref.	*Population investigated*	*Surrogate for traffic exposure*	*Health determinant*	*Findings*
51	A birth cohort of ~4000 children recruited during the second trimester of development from a series of communities, varying from rural areas to large cities, living in the northern, western, and central parts of The Netherlands	• Air pollution measurements of $PM_{2.5}$, NO_2 and soot at 40 individual sites • Regression models were developed to relate the annual average concentrations measured at the 40 sites with GIS variables.	• Parent completed questionnaires on wheezing, dry night-time cough, ear, nose, and throat infections, skin rash, allergic diseases • Physician-diagnosed asthma, bronchitis, influenza, and eczema at 2 years of age	Physician-diagnosed asthma, ear/nose/throat infections, and flu/serious colds indicated positive association some of which reached borderline statistical significance. No associations were observed for the other health outcomes analysed
52	3221 children from a school in a heavy traffic area and 2969 children from a school in a low traffic area	• Traffic counts	• Parent completed questionnaires	The freeway surrounding the children's school in the heavy traffic area may not be associated with an increased risk of respiratory symptoms
53	Children aged between 5 and 14 with emergency admissions for asthma (n = 1380), or all respiratory illnesses (n = 2131) and a control group (n = 5703)	• Proximity of residence to main roads • Traffic counts	• Hospital admissions for emergency asthma and/or all respiratory illnesses	Adjusted odds ratios of hospital admission for asthma and respiratory illness for children living within 150 m of a main road compared with those living further away were, respectively, 0.93 (95% CI 0.82 to 1.06) and 1.02 (95% CI 0.92 to 1.14). The study showed no association between risk of hospital admission for asthma or respiratory illness among children aged 5–14 and proxy markers of road traffic pollution

54	Population admitted to hospital with acute respiratory diseases for two study periods (1995–1997, 1997–2000)	• Air pollution measurements for PM_{10}, NO_2, SO_2, O_3, benzene, formaldehyde, and toluene	• Hospital admissions	Benzene had the strongest association with respiratory diseases for the total study period, the relative risk of an interquartile increase of benzene was 1.095 with 95% CI: 1.031–1.163; and 1.049 (0.990–1.112) for formaldehyde, 1.044 (1.000–1.090) for toluene, 1.064 (1.019–1.111) for NO_2, 1.043 (1.011–1.075) for SO_2, 0.990 (0.936–1.049) for O_3 and 1.022 (0.990–1.055) for PM_{10}. The study showed evidence for short-term respiratory health effects of traffic related air pollution
55	A random cohort of 5000 people (age 55–69 years) from 1986 to 1994	• Measured regional and urban background concentrations of black smoke and NO_2 and an indicator variable for living near major roads	• Cause specific mortality	Cardiopulmonary mortality was associated with living near a major road (relative risk 1.95, 95% CI 1.09–3.52) and, less consistently, with the estimated ambient background concentration (1.34, 0.68–2.64). The relative risk for living near a major road was 1.41 (0.94–2.12) for total deaths. Non-cardiopulmonary, non-lung cancer deaths were unrelated to air pollution (1.03, 0.54–1.96 for living near a major road). Long-term exposure to traffic-related air pollution may shorten life expectancy

Table 12 *(cont.)*

Ref.	*Population investigated*	*Surrogate for traffic exposure*	*Health determinant*	*Findings*
56	Hospital admissions in Rome between 1995 and 1997 for respiratory conditions for all ages, and for asthma among children (0 to 14 years)	Air pollution measurements of PM_{10}, NO_2, SO_2, O_3, and CO at five monitoring stations	• Daily counts of all respiratory conditions, acute respiratory infections including pneumonia, chronic obstructive pulmonary disease (COPD) and asthma among all ages. • Hospital admissions for respiratory conditions, acute respiratory infections, and asthma occurring among children (0 to 14 yrs) were also separately analysed	Total respiratory admissions were significantly associated with same-day level of NO_2 (2.5% increase per interquartile range (IQR) change, 22.3 $\mu g\ m^{-3}$) and CO (2.8% increase per IQR, 1.5 mg m^{-3}). No effect was found for particulate matter and SO_2, whereas O_3 was associated with admissions only among children (lag 1, 5.5% increase per IQR, 23.9 $\mu g\ m^{-3}$)
57	78 adult asthmatics residing within 2 km of an air quality monitoring site	• Air pollution measurements of particle number counts, PM_{10}, PM_2.5, PM1, NOx, SO_2, O3, and CO at the site	• Daily peak expiratory flow (PEF) measurements and symptom and medication diaries for 6 months	Daily mean number concentration of particles, but not particle mass (PM_{10}, $PM_{2.5\text{–}10}$, $PM_{2.5}$, PM_1), was negatively associated with daily PEF deviations. The strongest effect was seen for particles in the ultrafine range. However, the effect of ultrafine particles could not definitely be separated from other traffic generated pollutants, namely NO_X and CO. No associations were observed with respiratory symptoms or medication use

Subject Index

Accountability, 60
Acute (Short-term) Effects, 144
  impacts, 82
Aerosol particles, 118, 119
Aerosols, 119
Aggregate, 60
Aggregation, 49, 53
Air pollution, 41, 49, 56, 57, 61
  Quality, 124
  Quality Management Areas, 75
  Quality Objectives, 75
Aircraft, 81
  emissions characterization, 7
  engine emissions, 2, 4
  exhaust, 138
Airports, 137
Aliphatic and aromatic hydrocarbons, 81
  hydrocarbons, 84
Alkalinity, 93
American Cancer Society Study, 144, 147, 149
Antecedent dry period, 83
Anti-icers, 88
APHEA II study, 143
Aromatic hydrocarbons, 85
  pollutants, 81
Arsenic, 144
Aspect ratio, 105
Atmospheric deposition, 84
  models, 122
Average daily traffic density, 83
Aviation, 1

Backcast, 57
Backcasting, 50
Balancing ponds, 103
Benchmark, 46
Benchmarking, 50
Benzene, 144, 153, 155
Best Management Practices, 98
Bioaccumulation, 94
Biodegradation, 100, 103
Biofiltration, 98
  interactions, 103
Boating activities, 81
BOD, 93
Brundtland Commission, 35
1,3-Butadiene, 144

Cadmium, 84
Cancer, 143
Carbon dioxide, 13, 125
  monoxide, 141
Carcinogens, 143
Carrying capacity, 39
Catalytic converters, 92
Chemical adsorption, 99
  speciation, 93
Chemistry transport model, 122
Children, 151, 152, 154
Chlorides, 91
Chromium, 85, 92, 144
Chronic effect, 147
  pollution, 91
Cirrus clouds, 16
Climate, 1
  change, 9, 10

Cloud albedo, 118, 119
Cobalt, 92
CO, 135, 141, 153, 156
CO_2 emissions, 121, 124
Coarse particles, 135
COD, 91
COMEAP, 146
Commission for Integrated Transport, 67
Committee on the Medical Effects of Air Pollutants, 145
Complexation with dissolved organic matter, 93
Conceptualization, 37, 38, 44
Congestion, 41, 53, 57, 62
  charging, 76
Constructed pervious surfaces, 98
  wetlands, 98, 104
Container, 112
Contrails, 15
Control, 84
  options, 81
Cooling, 118, 119
Copper, 81, 84, 85
Cost Benefit, 50
Critical loads, 39

Degradation of water quality, 94
De-icing agent, 85
Demand, 125, 127
Departmental Performance Report, 60
Detention, 103
Diesel vehicles, 133
Dissolved oxygen, 87

Eco-efficiency, 41, 59
  diversity, 89
  effects, 91
Ecotoxicological impact, 93
EEA, 46, 57, 58, 60
Efficient technologies, 125
Emission inventories, 129
  reduction, 126
Emissions, 115, 120–127
  trading, 19
Energy efficiency, 125
Episodic discharges, 82
European Accession Countries, 131
  Environment Agency, 36, 46, 55, 57, 58
  Union, 131
Evaluation, 40, 50, 54, 55
  /monitoring, 49, 51
ExTra index, 137

Fecal coliforms, 81
Filter drains, 100
  strips, 98
Fine particulates, 123
First-flush shock load of pollutants, 88
Forecasting, 50
  /backcasting, 51, 48
Formaldehyde, 155
Freight traffic, 112, 115
Fuel Duty Escalator, 67
  switch, 126

Global atmosphere, 1
  Climate Models, 10
  warming, 25, 26, 32, 39, 44, 46, 88
  warming potential, 25, 32, 10
  Warming Potentials (GWPs), 32, 33
Glycols, 88
Green flight, 21
Greenhouse effect, 29, 116, 117
  gas, 25, 29, 32, 33, 116, 117, 119, 121, 122, 125
Grit trap, 104
Groundwaters, 82
Growth, 111, 115, 121, 126, 127
Gully pot, 87
GWP, 32

Harvard Six Cities Study, 147, 149
Health effects, 141
  metals, 91
Herbicides, 81, 85
Highway runoff, 84, 89
  surfaces, 82
Human health, 129
Hydraulic, 82
  conductivity, 106
Hydrogen, 25–30, 33
  economy, 25, 26, 32, 33

Hydrojetting, 86
Hyperresponsiveness, 153

Impervious surfaces, 82
Indicator framework, 36, 48, 49, 51
  system, 60, 63
Indices, 46–48
  aggregating, 46
Indirect greenhouse effect, 118, 119
Infiltration, 82, 99
  basins, 101
  trenches, 101, 98
Inland waterways, 81
INRETS, 137
Integrated Transport White Paper, 65, 67, 68
IPCC, 9, 16

Lagoons, 103
Latent variable, 45, 53
Lead, 84
Life-support, 38
Local Transport Delivery, 74
  Transport Plans, 75

Macroinvertebrates, 93
Macrophytes, 104
Marine transportation, 140
Measurements, 121
Metal speciation, 93
Methane, 13
Methyl tertiary butyl ether , 81, 84
Mitigation, 124, 126
Modal split, 113, 114
Monitoring, 36, 50, 54, 57–61
MTBE, 81, 84
Multi-Modal Studies, 73

National transport policy, 71
Nickel, 85, 144
Nitrates, 94
Nitrogen dioxide, 142, 152–156
  oxides, 13
NO_2, 142, 152–156
NO_x, 136, 138, 153

Oil and grease, 87
  consumption, 113
  interceptor, 104
Operational measures, 21
Operationalization, 37, 38, 44–46, 51, 62
Oxides of nitrogen, 5, 136, 138, 153
Oxidising power, 116, 118, 122
Ozone, 13, 143, 145, 153, 155, 156

PAH, 140
Particles, 5
Particulate matter, 142
Passenger traffic, 114
Peak expiratory flow, 151
Performance indicators, 37, 44, 46, 47, 50, 51, 55, 59, 60, 61
  measures, 36, 46, 55
pH, 93
Phosphates, 94
Platinum group elements, 92
PM_1, 151
PM_{10}, 135, 138, 142, 145, 149, 151–153, 155, 156
$PM_{2.5}$, 142, 149, 151, 152, 154, 156
Policy agenda, 35, 41
  analysis, 46, 61
Pollutant concentrations, 134
  removal capacity, 99
  removal efficiencies, 100
Polyaromatic hydrocarbons (PAH), 81, 84, 140
Polycyclic aromatic hydrocarbons, 81, 84, 140
Porous asphalt, 107
  paving, 107
Primary emissions, 130

Quality of life, 36, 47, 47, 52

Radiative forcing, 9, 10, 117, 118, 121, 122
Rail transport, 81, 88
Receiving waters, 82
Reflection, 116
Regenerative technologies, 126
Renewable, 39, 42, 43, 49
Residential road, 83

Respiratory illnesses, 152
Resuspension, 83, 133
Retention ponds, 103
Road pricing, 79
  schemes, 72
  sweeping, 86
  traffic, 111–113, 115, 125, 134
  transport, 81, 121, 131
Runoff storage times, 104
Runways, 88

Scattering, 116, 118
Scenarios, 115
SDS, 60, 61
  monitoring, 60
Sedimentation, 99, 103
  tanks, 103
Ship, 122, 124, 140
Snow, 85
Soakaways, 98, 101
Soot, 154
  particles, 17
Sources, 81, 82
Spillages, 81
Stormwater sweeping, 87
  system, 82
Strategic environmental assessment, 59
  Rail Authority, 72
Subsonic aviation, 11
Sulfate, 17
Sulfur dioxide, 140, 141, 145, 152, 155, 156
Summer weed, 84
Surface waters, 89
Sustainability, 67
  dimensions, 52
Sustainable, 40, 44
  development , 35–37, 40–44, 46, 47, 49, 55, 57, 61
  Development Strategy, 60
  Drainage System, 98
  mobility, 36
  systems, 82
  transport, 36–66
  transport index, 53
  transport systems, 36
  transportation, 35, 40, 55, 61
Swales, 98
System boundaries, 44, 63
  boundary, 51

Target, 42–44, 46, 50, 51, 60
Technologies, 125, 126
Ten year Plan for Transport, 68
TERM, 46, 57, 58, 59, 60
Titanium, 85
Toluene, 155
Total suspended solids, 83, 84
Toxicity, 94
  tests, 89
Traffic performance, 114
Transport, 40, 44, 45
  and Environment Reporting Mechanism, 57
  -derived pollutants, 82
  infrastructure, 65
  policy, 65, 66, 67
  related pollutants, 82
  sector, 25, 33
  system, 36, 37, 41, 43, 44, 50, 54, 55, 61, 62
  White Paper, 66
Treatment options, 82
Tropospheric ozone, 118, 122, 124
Tungsten, 92,
Tyre wear, 84

United Nationals Framework on Climate Change, 10
Urban motorway, 83

Vehicle emissions, 84
  Excise Duty, 72

Warming, 116, 118, 119
Wash-off, 82, 83
Water quality, 82
  vapour, 15
Wetlands, 103
Winter de-icing practices, 84
World Health Organisation, 142

Zinc, 81, 84, 85